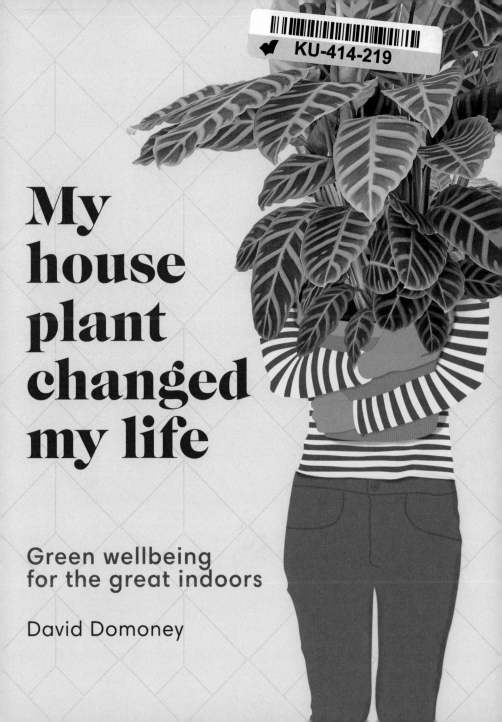

My house plant changed my life

Green wellbeing for the great indoors

David Domoney

CONTENTS

Get to know and love the 50 most life-changing houseplants, with all the care advice and expert tips you need to successfully grow your collection.

Top five...

Plants to calm and relax *p.50* • Plants to energize and enthuse *p.72* • Plants to cheer you up *p.94* • Plants to spark creativity *p.122*

Maidenhair fern
Adiantum raddianum
pp.34–35

Urn plant
Aechmea fasciata
pp.36–37

Barbados aloe
Aloe vera
pp.38–39

Tiger aloe
Aloe variegata
pp.40–41

Flamingo flower
Anthurium scherzerianum
pp.42–43

My house plant changed my life

Project Editor Holly Kyte

Project Art Editor Louise Brigenshaw

Senior Editor Alastair Laing

Art Editor Tessa Bindloss

Editorial Assistant Kiron Gill

Managing Editor Dawn Henderson

Managing Art Editor
Marianne Markham

Production Editor David Almond

Production Controller Luca Bazzoli

Jacket Designer Louise Brigenshaw

Jacket Co-ordinator Lucy Philpott

Art Director Maxine Pedliham

Publishing Director Katie Cowan

Illustrations Federica Gargiulo,
Mark Clifton

Photography Ruth Jenkinson

DK Delhi

Senior Art Editors Vikas Sachdeva,
Vinita Venogopal

Assistant Art Editors
Adhithi Priya, Ankita Das

Senior Managing Art Editor
Arunesh Talapatra

Senior DTP Designer Neeraj Bhatia

DTP Designers Vikram Singh,
Umesh Singh Rawat

Production Manager Pankaj Sharma

Pre-production Manager Sunil Sharma

First published in Great Britain in 2021
by Dorling Kindersley Limited
DK, One Embassy Gardens, 8 Viaduct
Gardens, London, SW11 7BW

A CIP catalogue record for this book
is available from the British Library.
ISBN: 978-0-2414-5851-8

Printed and bound in China

For the curious
www.dk.com

Bird's nest fern
Asplenium nidus
pp.44–45

Spider plant
Chlorophytum comosum
pp.56–57

Dendrobium
Dendrobium orchid
pp.66–67

King begonias
Begonia rex cultivars
pp.46–47

Natal lily
Clivia miniata
pp.58–59

Dumb cane
Dieffenbachia 'Tropic
Marianne'
pp.68–69

Zebra plant
Calathea zebrina
pp.48–49

Joseph's coat
Codiaeum variegatum hybrids
pp.60–61

Venus flytrap
Dionaea muscipula
pp.70–71

Hearts on a string
Ceropegia linearis
subsp. *woodii*
pp.52–53

Jade plant
Crassula ovata,
syn. *C. argentea*
pp.62–63

Corn plant
Draceana fragrans
pp.74–75

Parlour palm
Chamaedorea elegans
pp.54–55

String of pearls
Curio rowleyanus,
pp.64–65

Dragon tree
Dracaena marginata
pp.76–77

continued

Areca palm
Dypsis lutescens
pp.78–79

Guzmania
Guzmania cultivars
pp.88–89

Flaming Katy
Kalanchoe blossfeldiana
pp.100–101

Poinsettia
Euphorbia pulcherrima
pp.80–81

Velvet plant
Gynura aurantiaca
pp.90–91

Living stones
Lithops spp.
pp.102–103

Weeping fig
Ficus benjamina
pp.82–83

Amaryllis
Hippeastrum cultivars
pp.92–93

Prayer plant
Maranta leuconeura
'Fascinator Tricolor'
pp.104–105

Rubber plant
Ficus elastica
pp.84–85

Polka dot plant
Hypoestes phyllostachya
pp.96–97

Sensitive plant
Mimosa pudica
pp.106–107

Cape jasmine
Gardenia jasminoides
pp.86–87

Mexican hat plant
Kalanchoe daigremontiana
pp.98–99

Swiss cheese plant
Monstera deliciosa
pp.108–109

Sword fern
Nephrolepis exaltata
pp.110–111

Snake plant
Sansevieria trifasciata
pp.120–121

Pink quill
Tillandsia cyanea
pp.132–133

Moth orchid
Phalaenopsis
pp.112–113

Dwarf umbrella tree
Schefflera arboricola
'Compacta'
pp.124–125

Sky plant
Tillandsia spp.
pp.134–135

Heart-leaf philodendron
Philodendron cordatum
pp.114–115

Peace lily
Spathiphyllum 'Mauna Loa'
pp.126–127

Silver inch plant
Tradescantia zebrina
pp.136–137

Blue torch cactus
Pilosocereus azureus
pp.116–117

Madagascar jasmine
Stephanotis floribunda
pp.128–129

Flaming sword
Vriesea splendens
pp.138–139

African violet
Saintpaulia hybrids
pp.118–119

Cape primrose
Streptocarpus cultivars
pp.130–131

Spineless yucca
Yucca elephantipes
pp.140–141

AUTHOR'S PREFACE

I have no doubt that introducing indoor plants to your home will improve your life.

I also believe there is a houseplant that is just right for you and your home. In this book there are plants that can survive neglect, plants that love pampering, plants that flower like clockwork every season, and plants that look amazing every day of the year – and each one has a unique ability to improve your life, as well as its own individual personality that makes it special.

When we visit rooms that have no plants in them, they feel lifeless. A hotel room, a basement, a windowless office – with no greenery, they seem to suck the life out of us. Then think of a garden, with its lawn, its trees, its flowers, and how restored it makes you feel when you're surrounded by nature. Plants make us feel human and can have a significant impact on our physical and mental wellbeing.

I want you to feel a passion for bringing nature into your home and to experience the fun of nurturing and collecting houseplants. Because it's not just about caring for them; it's about appreciating the little things, like the beauty of a fresh new leaf or when they burst into flower, the amazement of your friends and family at what you have achieved, and the sheer enjoyment of the sensory experience whenever you find yourself marvelling at your plants' beauty. You'll be amazed at all the special benefits indoor plants can offer, and very soon these wondrous living organisms will become part of your family.

David Domoney
C Hort. FCI Hort.

Let houseplants into your life

Let houseplants into your life

HOUSEPLANTS AND MODERN LIVING

Human beings evolved as hunter-gatherers with an intimate relationship to nature. But thanks to increasingly rapid technological change, most of us, according to the World Health Organization, now spend 90 per cent of our time indoors, disconnected from plants and wildlife.

Nature on your windowsill

In recent decades, numerous scientific studies have provided a growing body of evidence to back up what experience seems to tell us: that engaging with plants and nature stimulates our senses, nourishes our minds, and has a beneficial effect on both our physical and mental wellbeing – to the extent that we should consider "nature deficit disorder" as a condition of our times.

Green-frame the problem

Of course, we can't all suddenly return to the wild and live off the land, but bringing nature indoors by growing houseplants is the perfect first step towards reconnecting with the natural world. Nurturing just one plant on a windowsill starts to re-establish an intimacy with nature. As your collection grows, your houseplants can create a living green frame through which to connect the inside with the outside, both literally and metaphorically: softening the harsh greys and straight lines of the built environment, drawing attention towards the attractive prospect of nature, and reframing your mindset to seek it out.

Bring the outside in
Create a "green frame" of plants around your window or office space to reconnect with nature.

Plant up to unplug

Modern living is increasingly dominated by the screen-based technologies of our phones, tablets, laptops, and televisions. While social media can provide a fantastic platform for sharing knowledge and inspiration, not least of the natural world, it is easy to become a slave to the algorithms.

Take a houseplant break

We all know the feeling of needing to disconnect and spend some time away from the screen. Caring for living plants provides that escape and washes away the pressures of the constantly connected world by replacing it with plants that don't make constant demands of us.

Looking at screens for long periods of time can also impact our physical health, leading to strained, dry eyes, blurred vision, and headaches. Our posture while engaging with technology can place strain on the spine, too, leading to aches and pains in the back and neck. So having a "houseplant break" to care for plants by watering, feeding, and generally appreciating them will not only clear our heads, but also give physical relief, refreshing both our minds and bodies.

Blue buzz, green relief

The presence of nature can help our brains to work better, with research showing that we associate the colour green with happiness, comfort, hope, and peace. This knowledge can help us combat the unseen effects of the technology all around us. The blue light emitted by computers, smartphones, and TVs, for example, can contribute to sleep issues by suppressing the sleep-promoting hormone melatonin. You can counteract this with foliage, as the green of your houseplants can instil feelings of tranquillity, safety, and calmness – the ideal conditions for a good night's sleep – and cleanse the air as you snooze.

HOUSEPLANTS AND MENTAL WELLBEING

I lecture all the time on the benefits of outdoor gardening to our mental health, but many would be surprised by just how good indoor gardening is, too – especially as we are spending more of our time indoors, with some people not having access to their own outside space.

I believe that indoor plants can make a practical and emotional contribution to our wellbeing. Our ever-changing urban modern societies, alongside ever-advancing technological developments, exert a constant pressure on humans to "keep up" with this fast-paced life. This pressure can trigger stress, low moods, low motivation, and low self-esteem as well as contribute to record levels of depression, PTSD, anxiety, loneliness, and OCD.

Caring for houseplants offers an enjoyable experience that can help in recovery. The appreciation of nature runs deep in our evolutionary make-up. For our hunter-gatherer ancestors, the bright colours of flowers signalled nutrition and a food source, triggering the "happy" hormone dopamine in the brain at the prospect of a reward. Even now, when we have food all year round, the blossoming of flowers still induces that same feel-good chemical, while becoming absorbed in caring for plants can provide a calming focus to help the mind recuperate.

Begonias fascinate with their intricate leaf patterns

Aloes surprise with unusual blooms

King begonias

Natural features
Intriguing begonias (*see pp.46–47*) and aloes (*see pp.40–41*) provide positive distractions and improve focus.

Tiger aloe

Focus & memory
The joy of seeing plants in flower or being fascinated by foliage can create positive distractions and spark happy memories.

• **Happy distractions** Being responsible for a living thing can break the cycle of negative thought patterns and replace them with positive ones, taking your focus away from your stresses and towards new life.

• **Aide-memoires** Plants have the power to evoke pleasant memories, either by sight or fragrance. They can change our mood in an instant, turning away sadness and replacing it with happy thoughts. Seeing a potted kalanchoe that your grandmother used to have may spark a treasured childhood memory; the scent of jasmine might be a reminder of your wedding day, a poinsettia suggests Christmas for many, while the scent of gardenia transports us to the summer holidays. Many such plants hold positive associations that have the power to make us smile.

Nurturing

Tending plants in your home or office will help you reconnect with nature.

• **The feel-good factor** Losing yourself in the care of plants frees your mind from any immediate stresses. This reduces levels

"The appreciation of nature runs deep in our evolutionary make-up**"**

of the stress hormone cortisol in your brain, resulting in calmer moods. Your mind is then focused on caring, which in turn releases dopamine, the hormone that makes us feel more positive and motivated.

• **Break the cycle** Caring for plants is absorbing and can soon become a regular hobby. This can help to break the cycle of low moods and depression, as forming positive new habits is a recognized element in cognitive behavioural therapy (CBT).

The spider plant sprouts plantlets that can be cut off

Allow the baby plants to root in jars of water

Once rooted, the baby plants can be potted up

Planting pride
Propagating and nurturing your own plants can bring an uplifting sense of achievement. The spider plant *(see pp. 56–57)* is a good choice to start with.

• **A sense of achievement** Plants can boost self-esteem. To see a living thing thrive thanks to your care and attention evokes a feeling of pride. And success in the form of propagation or seeing your plant flower releases serotonin, another hormone that lifts and stabilizes your mood.

• **A healthy relationship** There is evidence to show that plants can have a positive impact on those suffering from eating disorders, as their focus is shifted to caring for another living thing. The situation needs to be handled sensitively, but, with support, the final aim is to see the person growing their own salad, peppers, herbs, and mushrooms indoors, to help encourage a positive relationship with food.

Homeliness

Plants can enliven your home and bring comfort and positivity, too.

• **Visual therapy** Introducing plants to a room can benefit those who are living with post-traumatic stress disorder (PTSD), as changing the appearance of your space is encouraged as a form of therapy.

• **A breath of fresh air** Plants can prompt us to let light and air into a room. Opening windows encourages fresh airflow, while birdsong from outside will lift the mood and distract from negative thoughts.

• **The living home** Humans are living things. We are hardwired to be close to the

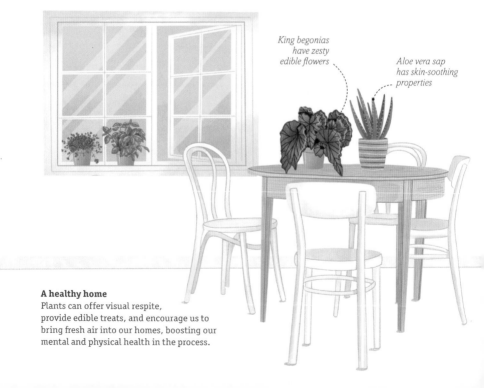

King begonias have zesty edible flowers

Aloe vera sap has skin-soothing properties

A healthy home
Plants can offer visual respite, provide edible treats, and encourage us to bring fresh air into our homes, boosting our mental and physical health in the process.

"Having **living things** in a house is what makes it a home"

natural environment, and dull, urban spaces devoid of plants can quickly become oppressive and induce feelings of anxiety. Having living things in a house is what makes it a home; family, pets, and plants all coexist to create that sense of cosiness. Adding growing plants and the natural colours of foliage and flowers can give a positive boost to your home and your mental state, reinforcing a feeling of security and familiarity and proving a happy sight to welcome you home.

• **Share the joy** Sharing your delight in houseplants with others adds another dimension to the magic of growing them. Keeping friends and family up to date with your successes will help you connect with them, as well as instil confidence in yourself, while making a gift of plants you have grown yourself shows love and care and will spread positivity to loved ones who see you nurturing life. Receiving a plant as a gift, meanwhile, is a living reminder of the person who gave it to you – every time you see it, it will lift your spirits.

Plant pets
As living, breathing organisms, plants can help bring a house alive, adding to the warmth and familiarity of your domestic space and even becoming a much-loved part of your family.

If you have pets, choose plants that aren't toxic when eaten or chewed

HOUSEPLANTS AND AIR QUALITY

Houseplants improve air quality in our homes by absorbing carbon dioxide and releasing oxygen, helping to achieve the right balance needed for breathing and optimal brain function. They can also help reduce the build-up of problematic toxins emitted by various household products.

In our daily lives we use numerous products that can leave residues in the air. Aerosols, cleaning products, gas stoves, open fires, paint, and cigarette smoke are just some of the sources of airborne toxins such as formaldehyde, benzene, xylene, and ammonia. Without proper ventilation, these toxins can build up in our homes and may cause headaches, nausea, low motivation, and irritation of the eyes and throat. They can also exacerbate skin problems and breathing difficulties such as asthma.

Get the air moving
Plants such as the p ill *(pp.132–133)* will appreciate an open window to boost airflow.

Plants can act as filters for these pollutants and help to neutralize them. A pioneering study in 1989 by the space agency NASA into the air-purifying abilities of certain houseplants has since been largely backed up by a 2007 Australian study, which found that keeping three or more large plants per person in an office can detoxify the air of volatile organic compounds (VOCs) by up to 75 per cent.

The average home would likely give different results, given the controlled environment in which these studies were carried out. But there are many things you can do to help your plants clean the air:

• The more plants, the greater the potential to clean the air. The exact number will depend on the size, maturity, and health of the plants, so just have as many as possible.
• Allow good air circulation by opening windows regularly and maximizing air movement (perhaps use a mechanized fan).
• Ensure your plants have the appropriate light, temperature, and humidity levels to optimize their health and efficacy.
• Help decrease the chemicals in your home by using low-emission products.

AIR PURIFICATION

The fundamental benefit of plants is that they oxygenize the air. When we breathe in oxygen, our bodies respire to produce carbon dioxide. Plants absorb that carbon dioxide as they photosynthesize and release oxygen, helping the cycle to repeat. In addition, plants can dispose of toxins in two different ways.

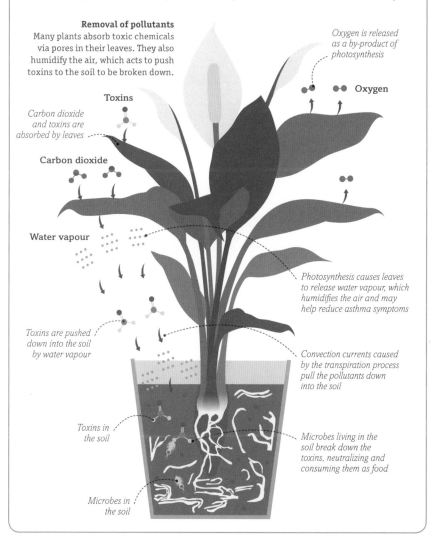

Removal of pollutants
Many plants absorb toxic chemicals via pores in their leaves. They also humidify the air, which acts to push toxins to the soil to be broken down.

Oxygen is released as a by-product of photosynthesis

Oxygen

Toxins

Carbon dioxide and toxins are absorbed by leaves

Carbon dioxide

Water vapour

Photosynthesis causes leaves to release water vapour, which humidifies the air and may help reduce asthma symptoms

Toxins are pushed down into the soil by water vapour

Convection currents caused by the transpiration process pull the pollutants down into the soil

Toxins in the soil

Microbes living in the soil break down the toxins, neutralizing and consuming them as food

Microbes in the soil

ENGAGING KIDS WITH HOUSEPLANTS

Getting children interested in plants will give them the best start in life, encouraging a positive interaction with living organisms and creating a healthy balance between artificial and natural environments. How you get them involved will depend on their age, but they'll soon be hooked.

Child-friendly chores
By the age of four, children will be able to start looking after houseplants such as aloes, jade plants, and living stones.

Babies and toddlers

These early years are magical and the perfect time for a child to have their first contact with indoor plants. For the very young, tactile plants such as the maidenhair fern and velvet plant offer a captivating sensory experience. And for the older children in this group, planting an amaryllis bulb and watching it grow will perfectly showcase the wonder and beauty of nature.

Young children

This age group is ready to take on watering duties and look after their own houseplants. Fun, safe varieties include hearts on a string, the fascinating living stones, aloes, and jade plant (though still make sure they don't taste the leaves) – all of which are visually arresting, yet easy for little ones to care for.

"Planting an amaryllis bulb will perfectly showcase the **wonder and beauty** of nature**"**

Pot up baby plants that form along the leaf edge

Make your own
For older children, the Mexican hat plant will be easy to propagate.

Tweens

Dexterity and mental development progress rapidly at this age, so introducing tweens to growing plants themselves from young plantlets or cuttings is a rewarding and educational pursuit. The fun polka dots of *Hypoestes phyllostachya* will pique their curiosity and the cuttings taken from it will root easily, giving the child a real sense of pride and independent achievement.

Young teens

Encourage this age group to grow fascinating plants on their windowsill. They are both old enough to be amazed by them and mature enough to carefully tend and nurture them. Varieties such as the Venus flytrap, the sensitive plant, and the prayer plant all have big personalities, as they move of their own accord, making them super interesting and great for showing off to friends.

Older teens

As teenagers move from childhood to adulthood and wish to distance themselves from their younger ways, plants are a great way to add low-maintenance, high-impact character to their rooms. As the bedroom becomes the "crib", cool architectural plants such as yuccas and palms add a bit of grown-up style, while striking plants such as the blue torch cactus make for excellent talking points. Likewise, chic orchids give real kudos.

The teen crib
Architectural plants such as the spineless yucca *(see pp.140–141)* are perfect for impressing your friends.

MOOD-ENHANCING PLANT COLOURS

Studies have shown that colour is a powerful stimulant that influences our emotions, with bright, saturated hues eliciting the strongest responses and paler tones having a calming effect. Plants showcase a rainbow of nature's colours, which can soothe, excite, or simply brighten our day.

Anthurium andraeanum, pp.42–43

Red

Although it is used in warning signs to alert us to danger, red is also a warm, energetic colour that can stimulate excitement and passion – hence why it's seen as an incredibly romantic colour. Vibrant, bold, and intense, this highly saturated primary colour makes a big statement and can really stir the emotions. A red houseplant is sure to ignite your senses.

Green

Humans can see more shades of green than any other colour, and when we think of it, we think of forests, woodland, and grass, which calms and refreshes us. Linked to good luck, safety, and health, green is a colour that helps us feel grounded and at one with nature, so bringing some greenery into your home is always a good idea.

The zingy green of this sword fern will refresh your space

Nephrolepis exaltata, pp.110–111

Pilosocereus azureus, pp.116–117

Blue
The soothing and tranquil effect of blue helps to aid relaxation and induce a feeling of serenity. Studies have shown that we associate the colour with clear skies and clean water, which have universal appeal. A blue plant is perfect for those moments of calm and mindfulness.

Saintpaulia hybrids, pp.118–119

Purple
Purple is a rare colour in nature, which makes it mysterious and intriguing when we see it in foliage and flowers. Due to its rarity, it's seen as a symbol of wealth and prosperity. Having purple plants in your home is also believed to spark creativity.

Yellow
Associated with optimism and happiness, yellow is a cheery, energizing colour that, like the sun, makes us feel warm. It is also thought to have a strong influence on the left side of the brain, which controls logic, perception, and deep thinking.

Codiaeum variegatum hybrids, pp.60–61

Orange
Like other warm colours such as red and yellow, fiery orange tones suggest enthusiasm and fun. This is a flashy, high-energy, attention-grabbing colour, while also being balanced, friendly, and inviting.

Clivia miniata, pp.58–59

White
Representing peace and purity, white is the most serene colour, bringing a sense of harmony to a room. Plants with white flowers or bracts and contrasting green foliage create a dazzling, fresh, and sophisticated feature.

Dendrobium orchid, pp.66–67

BUYING HOUSEPLANTS

Buying houseplants is an art. Always take a moment to consider where the plant will be placed, how it grows, how big it will get, and how to care for it. The plant may be in your home for years, so it's important to make the right decision. Here is my guide to buying wisely.

Check the label

Check that the plant label matches the plant you are buying. It's always best to keep a note of the common and botanical name, so you can research it later for more information and detailed care instructions.

Assess toxicity

If you have pets or young children who like to chew things, you need to be aware that some plants can be harmful if eaten or have sap that can be a skin irritant. Always check before buying that the plant is suitable for all in your household – I have highlighted the toxicity of each plant in this book.

Choose the right spot

If you have a location in your house earmarked for a new plant, keep a note of the conditions in that spot and check the label of your plant purchase to see if it is suitable. For example, if you have a full-sun location, it's not the best spot for ferns; if it's shady, it won't be ideal for a cactus.

Check for health

Check that the plant is in good health and keep an eye out for old plants. If the roots are growing out of the bottom of the pot, the plant is likely to be rootbound, which means it has outgrown its container. Give

Budding beauties
Buy your plant when almost entirely in bud so that it has plenty of energy left to flower at home.

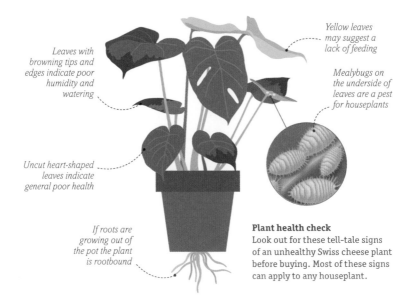

Leaves with browning tips and edges indicate poor humidity and watering

Yellow leaves may suggest a lack of feeding

Mealybugs on the underside of leaves are a pest for houseplants

Uncut heart-shaped leaves indicate general poor health

If roots are growing out of the pot the plant is rootbound

Plant health check
Look out for these tell-tale signs of an unhealthy Swiss cheese plant before buying. Most of these signs can apply to any houseplant.

---

the plant a thorough check over for the three Ds – are the branches or leaves dead, diseased, or damaged?

Buy in bud
It's always best to buy flowering plants when they are mostly in bud. That way, they will be ready to burst into flower as soon as you get them home.

Check the display
Ensure that the pot has drainage holes so that excess water can drain away; if it doesn't, it is important to repot it to prevent the roots sitting in water. For the same reason, check that the plant has not been left sitting in a tray of water while on display. *(See right.)*

Ask questions
Talk to the plant experts in the store and ask for advice. It's the best way to get to know your new plant and how to care for it.

"Check the plant for the three Ds – is it dead, diseased, or damaged?"

Lift the plant to check for overwatering

Beware root rot
If a plant continually drips water from the bottom of the pot when you lift it, it may have been sitting in water for too long, which can cause root rot.

WHERE TO POSITION YOUR PLANTS

Where you position your houseplant is perhaps the most important factor in determining its success. Some locations are perfect for certain varieties and hostile to others. And it's not just about light and warmth; there are many other influencing factors that make each spot unique.

Humidity When air around a plant is drier than in its cells, the plant can lose moisture through the pores in its leaves. The bathroom can be an ideal place for plants that like high humidity conditions.

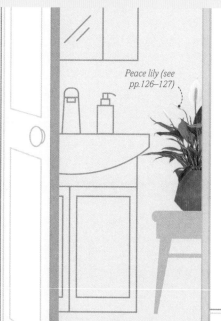

Peace lily (see pp.126–127)

Flaming Katy (see pp.100–101)

Draughts
Plants do not like constant changes in temperature. A draughty window or door position could stress a plant and during winter may kill it off.

Swiss cheese plant (see pp.108–109)

Plants for tricky spots For any really difficult positions like dark corners or spots that are either very humid or very dry, try tough, hardy houseplants that are forgiving, such as the Swiss cheese plant.

Air circulation Improving airflow helps cool high-up plants so they don't dry out so quickly, and generally helps reduce the chance of rot and disease from dampness.

Heat The heat from radiators can dry plants out and burn them, while temperature fluctuations can also cause distress, so choose tough, drought-loving plants such as cacti and succulents for hot spaces.

Heart-leaf philodendron (see pp.114–115)

Aim to position plants within easy reach, where you can access them for regular care

Pink quill (see pp.132–133)

Cape primrose (see pp.130–131)

Blue torch cactus (see pp.116–117)

Tiger aloe (see pp.40–41)

Light levels Generally, the darker the foliage, the darker the position a plant can tolerate, as this tends to mean a higher chlorophyll content to better extract light. Most plants, however, like a spot with bright yet indirect light.

CULTIVATING HOUSEPLANTS

Growing indoor plants and helping them flourish is a mixture of instinct, knowledge, and luck; I can help with the middle one, with these essential care tips on cultivating. Once you gain experience, you will be well on the way to creating your own indoor oasis.

Watering

Most plants are killed by kindness – from overwatering by being left to sit in water. Either water little and often or take your plants to the sink or bath and leave them to sit in the water for 20 minutes, then allow them to drain and return them to their positions.

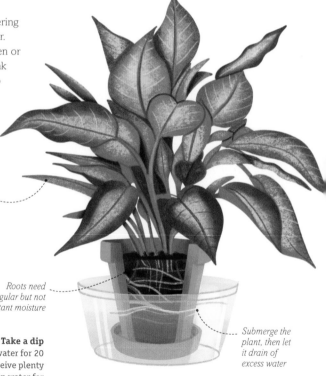

Leaves of a healthy, hydrated plant should be turgid and have good colour

Roots need regular but not constant moisture

Submerge the plant, then let it drain of excess water

Take a dip
Standing houseplants in water for 20 minutes will ensure they receive plenty of moisture without sitting in water for too long, which can lead to root rot.

Holiday solutions
Line your bath or shower with soaked towels and place your plants on top to ensure that they stay watered and well while you're away on holiday.

It's always best to water your plants with rainwater or distilled water if you can, because tap water contains additives such as chlorine, fluoride, and sodium. These are added for human consumption but can cause damage to some sensitive plants, building up on the foliage or in the soil, and leading to brown or wilted leaves and slow growth. Set up a rain-harvesting water butt for an eco-friendly water source.

Regular watering is of course essential, especially during the summer, so if you're going on holiday, put the plug in the bath or shower and line the bottom with old towels, adding water until the towels are sodden but not submerged. Then, stand all of your houseplants on the towels, leaving them until you return. The wet towels will provide the moisture and humidity the plants need.

> **"Clustering plants together allows them to create their own microclimate"**

Feeding
Feed your houseplants with liquid fertilizer or slow-release food for indoor plants during spring and summer, depending on their demands. Give plants a rest period between late autumn until early spring, with reduced watering and no plant food.

Humidity
Many houseplants – especially those with tropical origins – appreciate a humid environment. Regular misting is a simple and direct way to dampen moisture-loving plants, but for a lighter and more constant source of humidity, place the plant pot on a tray filled with gravel and water. This means that the pot isn't sitting directly in water, but as the moisture in the tray slowly evaporates, it creates a naturally humid environment.

These types of plants often grow better in groups. This is because each one releases water vapour into the air through transpiration, so clustering plants together allows them to share humidity and create their own microclimate.

Add liquid feed to water

Cleaning and inspecting

Keeping your plants' leaves free of dust stops the pores getting blocked, thereby helping them breathe more easily. Cleaning also allows plants greater access to the light that they need for efficient photosynthesis. Clean plants regularly by supporting each leaf in your palm, then wiping with a cloth or kitchen paper dampened with lukewarm water. Add a small squeeze of lemon to the water occasionally, to help remove residue build-up.

Keep it clean
Gently wiping leaves with a damp cloth to remove dust will help your houseplants to breathe.

Also, keep a regular eye out for insects or pests lurking on the stems and foliage, remembering to check the undersides of leaves. Treat any pests using appropriate care from your local garden centre and pull off any diseased or infected leaves.

Repotting

Repotting gives your houseplant room to grow and stay healthy. You should generally repot in spring, only when the plant's roots have outgrown the pot. Check the bottom of the pot to see if roots are coming through, insert your finger into the soil to see if it's solid with roots, or take the plant out of the pot to see if the roots are circling around the inside. If the answer to these checks is yes, the plant is rootbound and needs to be repotted.

Maidenhair ferns like good humidity and indirect light (see pp.34–35)

Right at home
Place moisture-loving plants in a tray of wet pebbles to help create a humid environment.

HOW TO REPOT YOUR HOUSEPLANT

Choose a container that is one or two sizes bigger than the one the plant is currently in, and ensure that the new pot has sufficient drainage holes to prevent the plant from sitting in water. Then follow these simple steps:

Soak the roots first

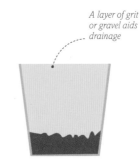

A layer of grit or gravel aids drainage

1 The day before you repot your houseplant, give it a good water so it's happier and easier to move.

2 Scatter gravel on the base of the new pot for drainage, then add suitable compost on top, leaving room for the plant.

Tease out roots to encourage growth

The roots will soon grow out into the new compost

3 Coax out the plant by gently tapping the bottom or squeezing the sides. Loosen the roots a little before planting it in the centre of the new pot and packing it down.

4 Add a little more potting mix around the plant, pressing it in until the plant is comfortably secure, then water well, and leave it to settle into its new home.

Houseplant
heroes

MAIDENHAIR FERN

Adiantum raddianum

The dainty foliage of this fern has a uniquely soothing quality. The beautifully fan-like leaves appear to float above the plant and, with the lightest breeze, flutter like confetti in the air. Everything about this plant is delicate in its detail.

I will make you smile

My bright, fresh, delicate foliage evokes a sense of fascination and simple wonder, refreshing the mind and inviting thoughts of woodland and nature. Blow on my foliage to see movement that will make you smile.

{ **"**My fluttering foliage brings a **calming motion"** }

Get to know me

"Maidenhair" refers to the goddess Venus, whose luscious locks my fluid foliage is said to resemble, while *Adiantum*, derives from the Greek meaning "unwetted", due to my moisture-repellent leaves.

Help me thrive

POSITION
Place in a bright spot away from direct sunlight. Likes medium to high humidity, and temperatures of 15–25°C (60–75°F) are ideal.

POTTING
Use multi-purpose, well-drained compost. Repot in spring, perhaps every 2 years, into a pot one size bigger to stop it becoming rootbound.

GROWTH RATE
Can reach heights of 45–60cm (18–24in), even as a young plant, but only gradually increases in width to its ultimate 80cm (32in).

CARE
Keep the compost moist with regular watering. Generally pest free but scale can be a problem, so catch it early and wipe it away with warm water.

I will calm you down

I love humidity and will be your best bathroom buddy, adding some fresh greenery to help you relax even more when you're having a soothing bath, and when you shower you'll be giving me a much-appreciated mist.

Rescue me!

I love moisture, but if I do dry out and look very sick, don't worry – I can be resurrected. Just cut off my dead, dry leaves and put me on a windowsill with indirect sunlight, ensuring my soil stays nice and moist, and there is a fair chance I will sprout again.

Be mindful of me

If I'm in a room other than a nice steamy bathroom or kitchen, try to maintain high humidity by misting me frequently, or place me in a wet pebble tray, which will increase the humidity when the water evaporates.

URN PLANT

Aechmea fasciata

There are few plants as spectacular as the Aechmea in flower. The tropical foliage dusted in silver creates a vase for the "jack-in-the-box" bloom borne on a long stem. This singular flower, which lasts for up to six months, looks like a firework with its spray of pink interspersed with effervescent blue.

{ **"**I'm an **air-purifying** tropical sensation**"** }

Help me thrive

 POSITION
Likes a warm, bright position out of direct sunlight, ideally 18–25°C (70–75°F). Needs a temperature of at least 25°C (75°F) to flower.

 POTTING
Be careful not to put it in an overly large pot as the roots might rot if they stay wet for too long. Using a free-draining orchid potting mix is best.

 GROWTH RATE
From a young plant it will take 2–3 years to reach its maximum size of 45cm (18in) tall by 60cm (24in) across.

 CARE
Keep the central urn formed by the leaves filled with fresh rainwater or distilled water. Also water the compost when it feels dry.

 I'll clean your air
As a member of the bromeliad family, I am different to normal houseplants and produce oxygen during the day and remove air pollutants in the night. When combined with normal foliage plants, we create 24-hour air purification.

 Let's make babies
I demonstrate the circle of life by dying after I have flowered. But before I go I will produce baby plants at my base to succeed me. When my top part has died fully, pull off my pups, getting as much root as possible, then pot them up.

 I need special watering
Please water the centre vase of my foliage, preferably with rain or distilled water. To stop it stagnating, replace with fresh water every month. Feed me by spraying half-strength fertilizer onto my leaves and roots every month in summer and every other month in winter.

Help me feel at home

I am an epiphyte, which means that in nature I grow attached to other plants and rocks, hence why I need to collect rainwater in my urn. At home you can grow me epiphytically by surrounding my root ball with sphagnum moss and placing me in a nook in a piece of driftwood, tying me in place using wire.

Watch I don't topple!

As a top-heavy plant I'd be happiest placed in a robust clay pot to stop me from toppling over. Although I don't need to be pruned to make me lighter, you can remove yellow or dying leaves and clean my leaves of dust by rinsing them.

BE AWARE:
The spines on my leaves can be a skin irritant.

BARBADOS ALOE

Aloe vera

Throughout history aloe vera has healed the sick and made the old feel youthful. Even Cleopatra is reputed to have used aloe as part of her daily skincare regime. Today, it's still vital to the health and beauty industry, and makes a fine architectural plant for any windowsill.

I'll clean your air

I can help alleviate ear, nose, and throat irritation by reducing levels of formaldehyde in the air, which is given off from everyday household detergent and soap products.

{ **"**I am a **living pharmacy** – the 'plant of immortality'**"** }

Help me thrive

 POSITION
Happy in direct sunlight or bright indirect light, and in temperatures of 20–25°C (70–80°F), though not dropping below 10°C (50°F).

 POTTING
Use a well-drained compost with plenty of grit. Repot when the plant is too crowded with suckers or remove them to repot and give away.

Care and share

If I am thriving, you may spot babies at my base, and you can spread the love by gifting these to friends and family. Once they reach 7cm (3in) tall, pull away each baby plant, with its roots, and pot them in free-draining compost.

 GROWTH RATE
This plant can reach heights of 60cm (24in). It will spread by suckering, but this can be limited by restricting the size of the container.

 CARE
Water every 3 weeks in spring and summer to keep leaves plump, letting the soil dry out in-between. In winter, once a month may be fine.

Be mindful of me

Please don't take all my leaves for your skin; leave me some so I can create energy to keep growing. Look out for soggy soft spots on my leaves – a sign you're giving me too much water. Remember, I'm a succulent, which means I store water in my leaves.

I will soothe your skin

Use the gel sap from inside my leaf to help soothe your sunburnt skin or as a natural skin moisturizer. I contain auxin and gibberellins – anti-inflammatory plant hormones that can help reduce scars, stretchmarks, acne, and signs of ageing.

Give me a holiday

In summer, I can go on holiday to your garden as long as I'm put in a warm, sunny spot. Acclimatize me over a week or two by placing me in the outdoor sun for a short spell each day, so I'm not overexposed. Remember to bring me back inside before the weather cools.

BE AWARE: I am mildly harmful if eaten, so keep me away from young children and pets.

TIGER ALOE

Aloe variegata

This was the first houseplant I ever bought, aged 10. With its bold, pointed foliage and radiating variegation, it seemed to have come from another planet and reminded me of the opening titles of *Doctor Who*! The succulent's easy-care nature means you'll get as much pride and wonder from growing it as I did.

I will cheer you up
Joy and happiness come from a feeling of accomplishment and success, which you are pretty much guaranteed by growing me. I am one of the easiest plants to grow because I store water in my leaves, which are waxy and coated, so I can survive periods of drought.

> **"I am low maintenance – perfect for a first-timer"**

Help me thrive

 POSITION
Best in direct sunlight or very bright indirect light. Happy in a dry spot as long as the temperature doesn't drop below 10°C (50°F).

 POTTING
Likes well-drained compost, but normal potting compost is fine as long as it's not overwatered. Repot every 2 years.

 GROWTH RATE
Will reach a height of about 20cm (8in) and spread by suckers or "pups", which can be removed and repotted (*see* Care and share).

CARE
The compost shouldn't stay wet for too long. Allow it to dry out before watering to make the soil slightly damp. Water less in winter.

Look out for my flowers

I will enrich your senses of sight and smell with my spring blooms, which are salmon pink and have a subtle sweet fragrance.

I open out into small tubular pink blooms

Get to know me

I am native to Namibia and South Africa and I have many common names, such as partridge breast aloe or tiger aloe because of my striking markings. In South Africa they call me *kanniedood*, meaning "cannot die".

Care and share

If you see pups appearing at my base, you can pull them away once the shoots are 7–10cm (3–4in) long and repot them, ready to pass on to friends and family – as such a low-maintenance plant to care for I make a perfect gift.

Be mindful of me

I am from a very dry area, so I do not like being in soggy compost. Please do not put me in a pot cover and let water sit in the bottom, because it rots my roots. I am better off being neglected than pampered.

BE AWARE: I can be mildly harmful if eaten, so keep me away from young children and pets.

FLAMINGO FLOWER

Anthurium andraeanum

The flamingo flower is a big fashion statement. The red heart-shaped flower and glossy green foliage evoke exotic vibrancy and beauty. The plant will bring an energizing touch of the tropics to your home and will work hard to improve air quality.

{ **"**I provide **bright, exotic blooms** all year round**"** }

Help me thrive

POSITION
Likes plenty of indirect light and room temperature as long as it doesn't drop below 15°C (60°F). Mist regularly to create high humidity.

POTTING
Repot every 2 years when the roots start to grow through the drainage holes or circle inside the pot. Use a loam-free potting mix.

GROWTH RATE
Reaches around 50cm (18in) tall and 30cm (12in) across. The showy bright-red spathes are produced intermittently throughout the year.

CARE
Water regularly to keep the soil moist during spring and summer. Water less in winter. Apply a houseplant liquid feed every 2–3 weeks.

I'll remove toxins

I can aid your attention and focus, reducing brain fog, headaches, and drowsiness, by removing xylene, a chemical compound that is released into the air by office and home computers or printer systems.

Get to know me

My flower is not as it seems, as the red plate is in fact a spathe, or modified leaf. My true flowers are minute and appear densely on the flower spike, or spadix, which is yellow or cream. The function of the spathe is actually to protect the flowers.

I'm a flower factory

As well as flowering throughout the year, my flowers last a long time – several weeks, in fact. I am not just a red-flowering plant, either; you can also find me in pink, white, mauve, green, and bi-coloured varieties.

I am symbolic

According to feng shui, I bring good luck in your relationships. Others say my red heart-shaped spathes signify long-lasting love and friendship, making me one of the most popular houseplants to gift to loved ones and friends.

BE AWARE: I am harmful if eaten, so keep me away from children and pets. My sap can also cause skin irritation.

BIRD'S NEST FERN

Asplenium nidus

This fern evokes the feel of a tropical forest. Its fresh green leaves erupt from the centre of the plant and spray upwards. It's perfect as a single specimen or in a display with other plants, adding texture and structure.

I'll clean your air
My large sculptural leaves mean there is more surface area for me to absorb carbon dioxide and convert it into oxygen to expel into the atmosphere, so I'm not only pretty but a practical addition to your home, too.

{ **"**Bathe in the **indoor forest** of my lush foliage**"** }

Help me thrive

 POSITION
A bright spot in indirect sunlight is best, though it will appreciate a bit of morning and/or evening sun. Keep it warm, not below 15°C (60°F).

 POTTING
It has a small root system but a big top half, so use a relatively small but heavy clay or ceramic pot for balance. A humus-rich compost is ideal.

 GROWTH RATE
Fronds can eventually reach 60cm (2ft) long but are quite slow growing – it can take between 4 and 8 years to reach its ultimate height.

 CARE
Likes humidity, so mist regularly and place on a tray of damp moss if possible. Compost should be damp but not wet, with no water sitting on the soil.

I will calm you down

My lush green colour is soothing and relaxing in an indoor environment – bringing the colour of nature inside evokes feelings of hope as new life grows. You can relax in the knowledge that I am also a very safe houseplant to have; I'm non-toxic to pets, or humans.

Help me feel at home

I am an epiphytic plant, meaning I gain moisture and nutrients from the air, which makes me a great houseplant for the more humid areas of your home, such as steamy bathrooms and kitchens.

Get to know me

My common name is bird's nest fern because the open rosette centre of my plant looks nest-like. On the undersides of my leaf I produce spores, which appear as brown lines running out horizontally to the mid-rib on my leaf.

Be mindful of me

My leaves are sensitive to chemicals, so please do not use leaf shine on me. Instead, just use tepid water and a damp cloth to clean the dust off my foliage.

KING BEGONIAS

Begonia rex cultivars

Words cannot describe the intricate merging of colours in the detailed patternation of this plant's foliage. You would be forgiven for thinking the leaves had been hand-painted by an artist, but this is all nature's work and occurs with every new leaf that unfurls.

I will cheer you up

My intricately patterned leaves are a tonic for the eyes. The fascination of my decoration can open your mind to the wonders of nature, putting a smile on your face. My hues of purple and red will also add some warmth to your home.

{ **"**Be **fascinated** by the beauty of my **leaves"** }

Help me thrive

 POSITION
Keep in a warm place, not below 15°C (60°F), where you can maintain high humidity. They prefer a bright spot but out of direct sunlight.

 POTTING
Use a free-draining potting mix and a large but shallow pot to allow the roots to spread. Divide and repot when it becomes rootbound.

 GROWTH RATE
They vary greatly in size according to the cultivar but are usually 30–50cm (12–18in) in height, which they'll achieve in 2–3 years.

 CARE
During the summer they will grow actively, so water and feed regularly, ensuring the compost stays moist. Keep drier in the winter.

My dainty flowers are pink and even edible

Get to know me

I come in many different varieties, including 'Escargot' with its snail-like spiral-shaped foliage, the deeply coloured 'Jurassic Rex' with its serrated leaf edge, the dark and moody 'Shadow King', and the bright and fruity variety called 'Cherry Mint'.

I'm a fussy drinker

With too little water I wilt quickly, and if left sitting in water, I can rot. Please water me little and often, without letting water sit on my leaves for too long. I'd also prefer it if you used rainwater or distilled water.

Taste my flowers

My delicate flowers are small, pink, and also edible, with a refreshing taste. They have a zesty lemony and peppery flavour, which comes from the oxalic acid in the petals. Do be aware, though, that oxalic acid can be harmful in large quantities.

Let's make babies

I am great at making baby plants. Just cut off one of my leaves, slice through some of my veins with a sharp knife, and lay me flat onto moist compost, weighing me down with stones to ensure I make good contact with the soil. Where a vein was cut, new plants will appear.

BE AWARE: I can be harmful if eaten, so keep me away from young children and pets.

ZEBRA PLANT

Calathea zebrina

This tropical beauty from the rainforests of Brazil is sure to engage both your mind and your senses, with its broad, lime-and-emerald striped leaves that also possess tactile qualities and a sensitivity to the circadian rhythms of day and night.

I'll clean your air

The large surface area of my leaves makes me a great air purifier, removing carbon dioxide from the air and releasing oxygen. And because my leaves also look so striking I am highly popular with interior designers.

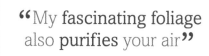

" My fascinating foliage also purifies your air "

Help me thrive

POSITION
Keep in a warm place, above 15°C (60°F), and away from draughts. Likes high humidity and a bright position but not direct sunlight.

POTTING
Repot every year or two, preferably in late spring, using a loam-based or loam-free compost, as they don't like being potbound.

GROWTH RATE
This plant can grow quite tall relatively quickly, reaching about 1m (3ft), and it will gradually bush out to 60cm (2ft) or more.

CARE
Loves moisture, so water frequently to keep the soil moist, leaving it drier in the winter. Feed monthly during the growing season.

Be mindful of me

I do not like rapid changes in temperature, so keep me away from draughts, especially in winter, and also heat sources such as fires and radiators, which will both rapidly change the temperature but also dry the air in the room.

Touch me

My zebra-striped leaves have a velvety feel to them because I have small hairs that cover my waxy leaves. I am a very tactile plant to engage with, and I can bring some height to your room as my oval leaves grow on top of long stalks.

Get to know me

I am known as a "living plant" because I close my leaves at night and reopen them in the morning. I have a small joint between my leaf and stem that enables light to move my stem, and as my leaves curl I can also make a soft rustling sound.

Help me feel at home

I get homesick for Brazil sometimes, so please place me on a tray of wet pebbles to keep the air around me moist, or give me some tropical rain occasionally by standing me in the shower under some lukewarm water – this will refresh me, water me, and wash my leaves.

PLANTS TO CALM AND RELAX

The power plants have to reduce the body's stress hormone cortisol is well established. Greens and whites calm us and soften our interiors by bringing nature indoors, while the sensory experiences of touch, sight, and scent enrich our spaces and evoke a feeling of tranquillity.

Dendrobium orchid ▼
Dendrobium

These towering beauties are adorned with flowers from the bottom to the top. Some of these orchids produce pure, tranquil, calming white blooms to help you destress, while their sweet, fruity scent will leave you feeling refreshed and renewed. They also possess valuable air-purifying abilities, which will make your space feel cleansed and fresh, and help you tackle your day with energy and enthusiasm. *See pp.66–67*

Barbados aloe ▶
Aloe vera

Not only is aloe vera an interesting structural plant for your interior, with its tall, fleshy spikes, it can also be used to soothe your skin if you feel as though you need pampering. Use the moisturising sap to invigorate, soothe, and cool your skin while your mind relaxes, so that your whole body experiences the benefits of nature. *See pp.38–39*

This delicate fern will rustle gently in the breeze

Peace lily ▾
Spathiphyllum 'Mauna Loa'
The peace lily has a pure white spathe that evokes a sense of harmony and tranquillity. Its purpose is to protect the true flowers, but it also contrasts beautifully with the glossy green foliage. In your home, these soothing colours will help ease those feelings of stress and strain. *See pp.126–27*

Maidenhair fern ▴
Adiantum raddianum
The soft, calming foliage of this fern flutters in the breeze like butterflies' wings. In a hanging basket, where it can flow freely over the pot, it will create a soothing feature as the bright yet relaxing green fronds sway and rustle above you. *See pp.34–35*

The spathe of the peace lily is calming

The weeping fig has a flowing habit that will soothe

◂ Weeping fig
Ficus benjamina
The fluid, arching stems of the weeping fig seem to flow like trickling water, emulating a waterfall captured in the moment. Standing like a tree in your home, it will bring the outside world inside to help you relax and feel connected to nature. The green foliage promotes feelings of calmness and safety, which in turn will give a sense of peacefulness. *See pp.82–83*

HEARTS ON A STRING

Ceropegia linearis subsp. *woodii*

You are going to fall in love with this little plant, dripping with baby heart-shaped leaves. The silver-dusted foliage runs along the length of its trailing branches, giving it a restful quality like a gentle waterfall. It's also easy to care for, so a great plant for gaining a sense of achievement.

{ **"** I'm a **living mobile** of cascading leaves **"** }

Let me cheer you up

I am known as the sweetheart vine, due to my romantic heart-shaped leaves. If given to a loved one, partner, parent, child, or friend, I will bring joy and remind the recipient of their connection to the one who kindly gifted me.

My quirky tubular flowers can attract pollinating insects

Help me thrive

 POSITION
Tolerant of different light levels but likes bright conditions best and is happy with some morning and/or evening sun. Keep above 10°C (50°F).

 POTTING
Use a free-draining compost (even cacti or succulent mixes are suitable) and don't plant in too big a pot, as it does well in crowded pots.

 GROWTH RATE
Can grow quickly and will look impressive after 3 to 5 years with 1m (3¼ft) stems. No need to prune unless it gets too big or untidy.

 CARE
Water when the soil is dry during the growing season – better to be cautious than to overwater. Don't let it sit in water, and keep it drier in winter.

Be mindful of me

I do best when crowded, so only repot me when I have run out of growing space. In the meantime, give me an occasional low-nitrogen feed to replenish my nutrients. This means I won't need to be repotted every year – one reason why I'm a great plant for beginners.

Help me feel at home

I have very long trailing foliage of well over 2m (6½ft) in length, so I am perfect for a high shelf in the corner of a room or in an indoor hanging basket, where I can create a feature point.

Look out for my flowers

While my foliage may be the main attraction, I do also bloom with unusual pinky-white flowers, which appear along my stems and can develop into cylindrical fruits.

Let's make babies

You can take cuttings from me by snipping off a piece of my vine and placing it into a jar filled with water, so you can see when roots appear. Once they have rooted, plant my babies up in their own pot.

PARLOUR PALM

Chamaedorea elegans

Made famous as a houseplant by the Victorians and placed in their parlours for exotic effect, this plant duly gained its common name, the parlour palm. The slender arching stems with a mid rib dressed with opposite fanning leaves of beautiful fronds make it one of the most popular indoor palms.

{ **"I'm resilient and will remove ammonia"** }

I'll remove toxins

I can help reduce respiratory irritation caused by exposure to airborne ammonia – a toxin released from bathroom-, surface-, and furniture-cleaning chemicals – as I absorb it into my leaves.

Help me thrive

 POSITION
This plant likes a spot in soft, indirect light, and the average household room temperature of 18°C (65°F) will suit it well.

 POTTING
Repot the plant every other year into a container that is 2.5–5cm (1–2in) larger than the original. Use an all-purpose compost.

 GROWTH RATE
These palms are slow growing, but in a few years they can reach their ultimate indoor size of 1m (3¼ft) tall and 50cm (1¾ft) wide.

 CARE
Water well in summer but let the top soil dry out between waterings. Water less in winter. Feed monthly with a balanced liquid fertilizer.

I have a secret

Not many people know this, but if I am well looked after and have plenty of light, as I get older I will flower, with masses of small golden globe-like flowers on my multiple stems.

*If I'm mature, yellow
flower sprays appear
on branching stems*

I am symbolic

My horticultural name is
Chamaedorea, which means
gift from the ground, from the
Latin *chamai* meaning "on the
ground" and *dorea* meaning
"gift". It suits me very well, as
I am believed to symbolise
the bounty of nature.

Watch I'm not too wet

It's better to underwater me
than overwater me, as I am
tolerant of some neglect but I
don't like sitting in water. So
stick your finger into the top
inch of my soil and if it's dry,
give me a thorough soak and
pour away any draining water.

Get to know me

I may look like one plant but actually I'm a
cluster of small plants growing together to
create a compact, bushy effect. This means all
of us are competing for the nutrients in the pot,
so when you repot me you may wish to divide
me to make more plants, but remember that
I won't have the same dense impact.

SPIDER PLANT

Chlorophytum comosum

This plant is one of the great survivors of the houseplant world. It can tolerate a lot of neglect and bounces back well when remembered. It's been a popular houseplant for decades, bringing bright foliage to empty windowsills with its beautiful arching leaves, which have fresh lime green and white variegation.

I'll remove toxins
I can reduce the risk of exposure to formaldehyde, which might be found in varying levels in wooden-furniture stains, varnishes, and polishes, adhesives, building materials, detergents, and cigarette smoke. I absorb the toxin into my leaves and release oxygen in its place.

> **"**I'm **hard to kill** and a great **air purifier"**

Help me thrive

 POSITION
Likes a bright spot, but avoid direct sunlight as this can burn its leaves. Grows best between 14°C (55°F) and 18°C (65°F) at a constant temperature.

POTTING
Transplant to a larger pot with an all-purpose houseplant compost when you start to see the roots emerge above the soil.

GROWTH RATE
These plants are rapid growers, producing plenty of sprouts. They can grow up to 60cm (2ft) tall and can get to 1m (3¼ft) wide.

 CARE
Allow the top of the soil to dry out before watering. Use a liquid houseplant feed every week or two in spring and summer.

I'll be your work companion

I can survive in artifical light, making me perfect for brightening up office spaces. Studies have also shown that a splash of greenery helps improve productivity and decrease stress.

I'm a good oxygenator

The white patches in variegated leaves lack chlorophyll, which gives a plant its green colour and is vital for photosynthesis. But I produce more chlorophyll than other variegated foliage plants, which means I release more oxygen into the atmosphere.

Enjoy my leaves!

My horticultural name, *comosum*, means to have long or abundant hair, which suits my slender, sword-like leaves. In fact, my leaves are also edible, so I am a particularly safe plant to have around pets and young children.

Let's make babies

I am also known as the "hen-and-chickens" plant, as I sprout many babies on long stalks, which trail down. These little plantlets can be cut off and potted up as new plants or left hanging down to give me a more interesting shape.

NATAL LILY

Clivia miniata

I bought a Clivia for my parents 15 years ago, and every year my dad calls me to tell me it's flowering again. I'm not surprised at his excitement, because this plant has the most spectacular flowers and blooms annually. The real beauty is the colour combination: the vibrant tangerine flowers almost glow against the deep emerald foliage.

> **"** I'll bring **happiness** with my **bright** blooms **"**

I am symbolic

My blooms are long-lasting, so I have come to signify long life. I became a popular indoor plant in the Victorian era and inside the palaces of the last Chinese imperial dynasty because of my symbolic longevity. I don't mean to show off, but in Changchun I am the city's emblem.

Help me thrive

 POSITION
To encourage flowering, place in a well-lit, cool conservatory or any bright, unheated room from late autumn until buds form in early spring.

 POTTING
Doesn't like being disturbed, so only repot by one size up, using a well-drained, loam-based compost, when it's bursting from its container.

 GROWTH RATE
This plant can grow up to 45cm (1½ft) tall and 30cm (1ft) wide. It is slow growing, so it may take 2–5 years to reach maturity.

 CARE
Water freely, letting the top of the soil dry out before watering again, and feed weekly in spring and summer. In the winter months, water sparingly.

Help me feel at home

I am a distinctive and beautiful foliage plant. My dark-green, sword-like leaves will stay blemish free if I'm not put into a room that is too hot or too dry. When the temperature drops in the winter, stand me on a tray of wet pebbles to make the atmosphere a little more humid.

Be mindful of me

Please feed me once a month in spring and summer. When I'm thirsty I prefer rainwater or distilled water because I am sensitive to chlorine damage. I flower best when I am rootbound in my container, so keep me in it for up to four years before repotting. I do get top heavy, so choose a heavy pot to keep me stable.

Get to know me

I am native to South Africa and was first identified in KwaZulu-Natal in the early 1850s. I get my genus name from Lady Florentina Clive, who first cultivated me in the UK, and *miniata* means "bright red" – referring to my vibrant flowers, though they can be found in a variety colours.

BE AWARE: I can be harmful if eaten, so keep me away from children and pets.

JOSEPH'S COAT

Codiaeum variegatum hybrids

There are few plants that come anywhere near the mood-enhancing beauty of this one. It's not just the outrageous fusion of foliage colour, but the way the veins are spectacularly enhanced by their contrasting colours of yellows, red, oranges, greens, and even purples. You could pluck off a leaf and frame it on the wall – it really is that special.

Help me feel at home

I'm happiest in a room with a stable temperature and no draughts, otherwise I risk dropping my leaves. I can grow bushy, so a spot with room to grow is best, but if I get too big, you can prune me back as much as necessary.

{ **"**My **technicolour dazzle** will lift your spirits**"** }

Help me thrive

POSITION
Bright, indirect light keeps the colours vibrant, so by a south-west, south-east or east-facing window is ideal, in temperatures of at least 15°C (60°F).

POTTING
Keep in the same pot to control its size, or repot in spring for growth, using a pot one size bigger than the original and a free-draining compost.

GROWTH RATE
Indoor plants tend to grow to 60cm (2ft) tall and are slow compared to plants grown in a greenhouse, which can reach 2–3m (6½–10ft).

CARE
In spring and summer, water to keep the soil moist and mist daily, then feed every 2 weeks. Water less in winter so the soil is moist, not wet.

Get to know me

I have brothers and sisters with many different leaf shapes – they can be long and thin, or form curls and corkscrews. They also come in many colour combinations, such as fiery reds or deep purples, so you can have fun collecting them. We are all unique in our patterns and colours, so when you choose one of us you'll know we're one of a kind.

Admire my fabulous foliage

My amazing foliage will brighten up your day just by being seen – I cannot be missed. Sometimes my leaves even change colour as I mature, which is why I symbolize change. With my foliage stealing the show, my flowers aren't always noticeable because they are discreet white or yellow blooms.

Rescue me!

If I get too spindly and leggy, you can give me a trim to encourage new and bushy growth. Cut me back to 12cm (5in) from the top of my pot and dust some powdered charcoal over the cut ends to stop my sap running out.

BE AWARE: My sap can cause skin irritation and my leaves are harmful if eaten.

JADE PLANT

Crassula ovata, **syn.** *C. argentea*

My gran had one of these for years – I remember seeing it as a child. I inherited it, took cuttings from it, and it's been in my life as long as I can remember. The jade plant has oval leaves and grows like a small tree. The traditional form is deep green, but there are some more colourful cultivars available, such as 'Hummel's Sunset', pictured here.

I can boost your energy

Feng shui experts believe I aid mental wellbeing by nourishing *chi* – the bright, refreshing, and uplifting energy. Placing me in eastern positions is said to help family harmony and health, while western locations are believed to promote creativity.

I form clusters of pinky-white stars as winter blossom

> **❝** I'll bring **luck** and **luscious leaves** to your home **❞**

Help me thrive

 POSITION
Will flourish with plenty of light – it will be happy if it gets a few hours of sunshine every day in a minimum temperature of 5°C (40°F).

 POTTING
When the plant becomes rootbound repot in spring or summer – use a heavy container, as they can get top heavy, and a well-drained compost.

 GROWTH RATE
These mini-tree-like plants grow slowly. They can reach up to 1m (3ft) tall and wide, taking 10–15 years to reach their ultimate size.

CARE
Feed monthly and water often during spring and summer so the soil stays moist, but water sparingly during the winter months.

I am symbolic

I am also sometimes known as the "friendship plant" or "money plant", as I am said to symbolize great friendship, luck, and prosperity. I am also a popular choice as a wedding gift, as I can live for many, many years and will stay with you throughout your whole life, bringing good luck.

Look out for my flowers

Another of my names is the "flowering jade plant" because in suitable conditions I can produce white star-shaped flowers in winter. They complement my glossy leaves perfectly and will also provide a gentle aroma. But don't expect me to flower every year!

Let's make babies

If one of my leaves or branches breaks off, you can make a baby plant from it. Just cut the end cleanly and sit the base in water and watch it root. Once my babies have rooted, pot them up in a well-drained succulent mix.

Get to know me

I have a similar appearance to a bonsai tree, as my habit means I grow like a miniature tree with a thick, sturdy trunk and branches. However, I am a succulent, meaning I store water in my foliage, so I am considerably easier to keep than my bonsai friends.

BE AWARE: I can be harmful if eaten, so keep me away from young children and pets.

STRING OF PEARLS

Curio rowleyanus

You will love this remarkable succulent for its easy-going nature and unusual pearl-like foliage. As they grow, the stems form a delicate cascade, which has a restful effect on the mind, much like trickling water. The aromatic flowers will further stimulate your senses.

> **"** I'm so **easy to share** with the people you love **"**

Get to know me

My green pearls are an odd type of leaf whose shape reduces evaporation, helping me survive in the desert. Each "pearl" has a transparent slit that acts as a window, allowing light to flood inside and help power my photosynthesis.

Help me thrive

 POSITION
Place in a bright spot out of direct sunlight. Grows best at 18–24°C (64–75°F) but it will tolerate short periods at 10°C (50°F).

 POTTING
Use cactus compost or a 50:50 mix of 4mm grit and loam-based compost. Repot in spring in dry compost, and water after 2 weeks.

 GROWTH RATE
In the right conditions, this plant can grow 20–25cm (8–10in) per year, reaching an ultimate length of 1m (3ft).

 CARE
Water every 2 weeks in spring and summer, and not at all in autumn and winter (though don't forget to say "hi"!).

Be mindful of me

Give me an energy boost by adding a high-potash fertilizer to my water two or three times a year. Pull off any old, shrivelled leaves and trim back stems in spring, as necessary to keep its size and appearance in check.

I'll delight your senses

With the right care, I'll provide aromatherapy in summer, when I produce lots of white flowers with a pleasant, spicy scent. Take a sniff and enjoy the mood-enhancing power of smell

Rescue me!

Watch out in summer in case my pearls are starting to look a little scorched from too much direct sunlight. Just move me temporarily to a shadier spot.

Let's make babies

If you want to care for and share my offspring, snip off a stem in early summer, cut it into pieces, and lay them on fresh soil with a thin layer of sand in a pot. Watch till tiny roots appear before watering.

BE AWARE: I can be harmful if eaten, so keep me away from young children and pets.

DENDROBIUM

Dendrobium orchid

We had a stunning white one of these in our bathroom for years. It is such a beautiful flowering plant, with a vertical stem blessed with alternate green leaves and plenty of flowers all the way up its stem. From the base to the top, the whole plant is in bloom, and they can have wonderful fragrances, too.

I'll clean your air

My leaves may remove toluene toxins from the air, which are released by paint, glue, and nail polish, and may have a detrimental effect on the central nervous system. I also respire and release oxygen at night, so I am great for your bedside table, where you can benefit from the oxygen while you sleep.

"I'm a **flower tower** with an **uplifting scent**"

Help me thrive

 POSITION
Keep in full light in autumn and winter and light shade in spring and summer. It's best suited to temperatures between 18°C (65°F) and 30°C (85°F).

 POTTING
Tolerates being rootbound, and the pot should have plenty of drainage holes. Prefers an aerated potting mix, with good drainage.

 GROWTH RATE
Can grow up to 60cm (2ft) tall and 30cm (1ft) wide. It's a top-heavy plant and will grow best if staked so the cane grows upwards nicely.

CARE
In the summer, mist daily and water when the top of the soil feels dry. Feed every third time you water. Keep the plant drier in winter.

I stimulate multiple senses

My pretty flowers will not only be a refreshing treat for your eyes, they also have a delicious sweet and fruity scent like that of strawberries and raspberries. This lovely aroma will calm you and lift your mood.

I bloom in many colours
I am available in a range of flower colours, from white and green to double pink shades, gold and orange, purple and white, and many more – you can build quite the multicoloured collection.

Help me feel at home
In my natural habitat I am an epiphyte, growing on other plants where I get plenty of free-flowing air. Please make me feel at home by placing me in a spot where I can get access to good airflow, such as a hallway, landing, or bathroom.

Get to know me
I have a big family – there are more than 1,000 species of me that can be found all around Asia, the Pacific Islands, and Australia. My family are native to many different habitats, from the exotic tropics to mountain forests and even dry desert climates.

DUMB CANE

Dieffenbachia **'Tropic Marianne'**

This plant has a lot going for it: the
soothingly soft, fresh shades of its green
foliage and its air-purifying qualities.
Little wonder it continues to be one
of the most popular houseplants to
grow – just don't try to eat it!

I'll remove toxins

I use my leaves to absorb
xylene from the air – a sweet-
smelling hydrocarbon used in
paints, polishes, and solvents.
By doing this, I'll help to
alleviate irritability and
symptoms of depression
caused by long-term exposure
to the toxin.

" My lusciuos leaves will
take toxins from the air **"**

Help me thrive

 POSITION
A warm room with
indirect sunlight is ideal;
this is especially important
in spring and summer,
as tender leaves will be
scorched by direct light.

POTTING
Repot every year
or two in spring into a
slightly larger container.
Use a fast-draining
compost, adding gravel or
pebbles to help water flow.

 GROWTH RATE
Can grow up to
80cm (2½ft) tall and
60cm (2ft) wide, and
won't need any pruning
apart from the removal
of yellow leaves.

CARE
In summer, water
to keep the soil moist
(maybe even twice a
week) and feed monthly.
In winter, water sparingly
and place in good light.

Help me feel at home

I like humid environments, so
place me on a wet pebble tray
or use a humidifier to keep
me happy. I'm also at my best
with other plants. I thrive
when displayed or potted up
alongside others, because
when we're together, humidity
levels are higher.

I will keep you calm

You humans can see more green shades than most other colours, and green is a good colour to calm and soothe the brain. I have it in spades, so let me relax you with my refreshing shades.

Love my leaves

The ribbons of dark green, light green, and cream on my paddle-shaped leaves make me a handsome foliage houseplant. Turn me frequently so all of my leaves get access to bright, indirect light, and avoid direct light, as this will burn them. This is why I will thrive better than most in an office with limited natural light.

BE AWARE: My sap is harmful if eaten, so keep me away from children and pets.

VENUS FLYTRAP

Dionaea muscipula

Few plants have had pencils probed into them as often as this one. The Venus flytrap is every school child's dream. And the magnificent blood-red throat, green stems, and outer lashes are every fly's nightmare. A plant with such a strong personality is sure to engage your mind.

I will fascinate you

There are few plants that will fill you with more awe than I will when you see me eating my prey. The hair-like structures on my leaves can sense the presence of insects, and I respond by snapping my jaws shut. I then secrete digestive juices to consume them.

{ **"I'm a lovable monster who'll eat your flies!"** }

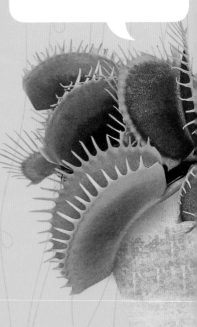

Help me thrive

 POSITION
Does best in a brightly lit spot, but can tolerate partial shade. Avoid putting them in direct sunlight as they may burn up.

 POTTING
Can be grown in a pot with drainage holes or in a terrarium with gravel underneath. A mix of sand and sphagnum moss is best for drainage.

 GROWTH RATE
These are slow-growing plants, reaching a maximum height of only 30cm (1ft) and a width of 15cm (6in) over a period of several years.

 CARE
In summer, stand in a saucer of rainwater or distilled water. In winter keep the soil moist. They don't need fertilizer as they get nutrients from flies.

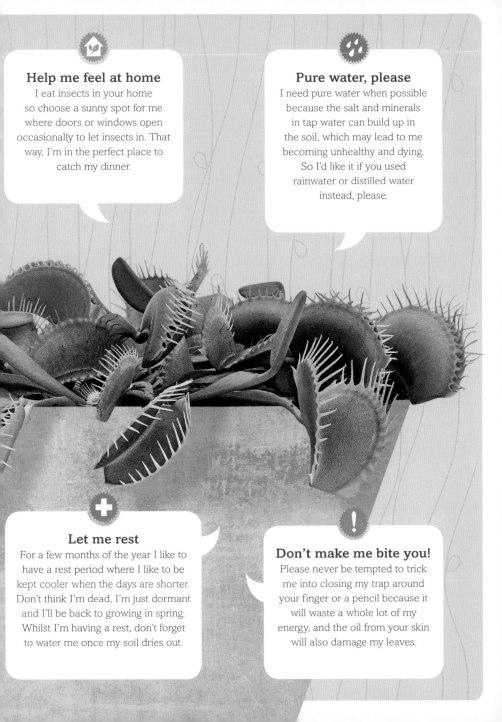

Help me feel at home
I eat insects in your home so choose a sunny spot for me where doors or windows open occasionally to let insects in. That way, I'm in the perfect place to catch my dinner.

Pure water, please
I need pure water when possible because the salt and minerals in tap water can build up in the soil, which may lead to me becoming unhealthy and dying. So I'd like it if you used rainwater or distilled water instead, please.

Let me rest
For a few months of the year I like to have a rest period where I like to be kept cooler when the days are shorter. Don't think I'm dead, I'm just dormant and I'll be back to growing in spring. Whilst I'm having a rest, don't forget to water me once my soil dries out.

Don't make me bite you!
Please never be tempted to trick me into closing my trap around your finger or a pencil because it will waste a whole lot of my energy, and the oil from your skin will also damage my leaves.

PLANTS TO ENERGIZE AND ENTHUSE

The sight of bright, engaging colours can fire up our emotions and boost our energy. Reds, oranges, and yellows are lively and uplifting, and when mixed with the dynamic shapes of the more flamboyant houseplants, they can stimulate while connecting us with nature.

Urn plant ▶
Aechmea fasciata

This plant is like a firework captured mid-eruption, with the magical contrast of the silver and green foliage against the central pink bracts, which are joined by small purple flowers sprouting from the core. The explosion of colour will fill you with excitement and exhilaration, and add some natural colour to your interior. *See pp.36–37*

Flamingo flower ▼
Anthurium andraeanum

The recognizable bracts of the flamingo flower are bright red – a colour of passion and action that stirs strong emotions. Having this vibrant bloom in the house will stimulate and energize. Of course, it's also the colour of romance, and the bracts are heart-shaped, so it's a symbolic gift for that special someone. *See pp.42–43*

The colour explosion of the urn plant will fire up the senses

Joseph's coat ▾
Codiaeum variegatum
The vibrant fusion of fiery colours on the leaves of this plant will inspire energy and vitality as well as adding a really colourful statement piece to any room in the house. Yellows, oranges, reds, and greens mix in bold patterns to encourage a get-up-and-go attitude, while the variations of the leaf shapes add even more interest and wonder. *See pp.60–61*

The red, star-shaped guzmania is bursting with positive energy

Guzmania ▴
Guzmania varieties
The thrust of energy provided by the vibrantly coloured, star-shaped bracts of this plant can't help but make you feel upbeat and enthused. Its flower-like sprays stick around for such a long time, and just before they say goodbye they find the energy to produce a baby to continue the family – that will really spark a feeling of magic and excitement. *See pp.88–89*

The woven neon bracts of the pink quill are pure fun

Pink quill ▸
Tillandsia cyanea
The electric pink shades of the quill, as well as its pleated texture as it splays out from the arching leaves, will add real excitement to your decor, while the purple flowers offer extra charm. There's an almost childlike quality to this fun, zingy plant that will make you feel playful, joyful, and young. *See pp.132–33*

CORN PLANT

Dracaena fragrans

This is one of the heavyweights of the houseplant world – and a great guzzler of toxins, too. Its tree-like trunks and clusters of bi-coloured foliage, growing to head height, mean it has great impact in any home. I bought them in their thousands when I worked in the trade.

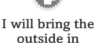

I will bring the outside in

I am a big player in softening the harshness of a room. As I can grow up to 2m (6½ft) tall, I will be a striking presence when stood in the corner of your room, where I will bring a touch of nature indoors. My living foliage will make your house feel alive and more like a home.

{ **"I'll be a striking natural presence in your room"** }

Pink-white starburst flowers may shoot from my crown

Help me thrive

POSITION
Needs filtered light, as direct sunlight will burn the plant, but too little light may cause the leaves to lose their stripes. Keep at around 18°C (65°F).

POTTING
Repot every year or two in spring into a slightly larger container with fresh potting soil, but don't pack it too much so as to retain good drainage.

My flowers are a rare treat
Although it's a rare sight when I'm grown indoors, I can flower when I mature. When my white flowers appear, you will realize why my name is *fragrans* – their scent is very strong and sweet. Some people choose to pinch off my flower buds so I can put more energy into my foliage.

GROWTH RATE
Can grow up to 2m (6½ft) tall and over 1m (3¼ft) wide. If the plant grows taller than you want, you can cut off the top of the canes.

CARE
Allow the soil to dry out a little before watering in spring and summer. Reduce watering in winter. In the growing season, fertilize frequently.

I'll remove toxins

I'm very effective at ridding the air of toxins. My leaves can help reduce nausea and confusion by absorbing xylene in the atmosphere, which is caused by paint, thinners, and other home-decorating products.

I'm easy to look after

As long as I get the basics – filtered sunlight, moist soil, and the occasional feed while I'm growing in spring and summer – I don't demand a lot of care. I am a perfect houseplant for beginners and for little work I'll reward you with my palm-tree-like presence.

Let's make babies

I am easy to propagate: simply cut my top off just below the leaf line and plant the cutting in some soil or in a vase with fresh water. Then, when the roots grow to 2.5cm (1in), plant it into a houseplant potting mix with good drainage.

BE AWARE: I can be harmful to pets if eaten, so keep me away from cats and dogs.

DRAGON TREE

Dracaena marginata

This is a beautiful small indoor tree that matches its looks with its strength to survive and ability to remove health-sapping toxins. I have always enjoyed the subtle variegation of the foliage, whose spiky tufts give the plant a pleasant palm-like structure.

I will boost your mood

My palm-like appearance will transport you somewhere exotic, making you feel cheered and relaxed. My foliage arches and hangs over with age, making me look more like a weeping willow than a palm and adding soothing shape and structure to your room.

{ **"My bold presence** will take you to the tropics**"** }

Help me thrive

 POSITION
A bright location with indirect sunlight is best for this plant, as direct light will damage its leaves.

 POTTING
Repot every other spring into a new container that is slightly bigger than the original. Use an all-purpose potting mix.

 GROWTH RATE
These are slow growers, taking about 10 years to grow to full size. When they do, they can reach up to 3m (10ft) tall.

 CARE
When in growth, water to keep the soil moist and give a monthly half-strength, slow-release feed. Water much less in winter, letting the soil dry out.

Get to know me

The gum-like sap that comes out of my stems is red like dragon's blood, hence my name. In medieval times, this gum was used in alchemy and the magical arts, but today it has a more practical use as a varnish for fine wood.

I'll remove toxins

I can help you maintain a healthy immune system by filtering benzene from the air. This hydrocarbon comes from the emissions of petrol-fuelled cars and cigarette smoke and is associated with various adverse health effects.

Love my leaves

You can also find me in other colour variegations, including tri-colour specimens, which have pink, green, and yellow leaves. My old leaves can be removed from the base of the trunk and you can cut back my stems to keep me at the size you like.

Be mindful of me

I am an easy plant to keep and have the ability to be drought tolerant if you forget to water me. Please look after my leaves by cleaning the dust off them with a wet cloth so I can breathe more easily.

BE AWARE: I am harmful to pets if eaten, so keep me away from cats and dogs.

ARECA PALM

Dypsis lutescens

The first areca palm I owned grew to head height and arms' width. It transformed my student bedsit with its glorious arching fronds of light-green foliage and contrasting golden stems. Don't underestimate the powerful presence of this graceful Madagascar palm – nor its air-purifying ability.

I'll clean your air

I help with healthy breathing functions because, through photosynthesis, my leaves are very effective at extracting carbon dioxide and can assist with filtering levels of carbon monoxide from the air and replacing them with oxygen.

If I bloom, my bright yellow sprays appear just under my leaves

{ **❝**I will bring **structural elegance** and **clean air❞** }

Help me thrive

 POSITION
Grows well in a bright, light room out of direct sunlight. Don't let the temperature drop below 15°C (60°F), as this may discolour the leaves.

 POTTING
Repot every 3 years in spring into the next pot size up. Use a free-draining potting mix, as these plants hate overly damp conditions.

 GROWTH RATE
Areca palms usually grow up to 25cm (10in) a year, and can ultimately reach a height of 2.1m (7ft).

 CARE
When in growth, water to keep the soil moist but not wet, and give a monthly balanced liquid feed. Water infrequently during winter.

I'm quite a looker

I'm also known as the "butterfly palm" because my structure resembles the shape of a butterfly's wings. The strong stems at my base support my tropical fronds, which arch upwards, reach a peak, and then arch gracefully downwards like a fountain.

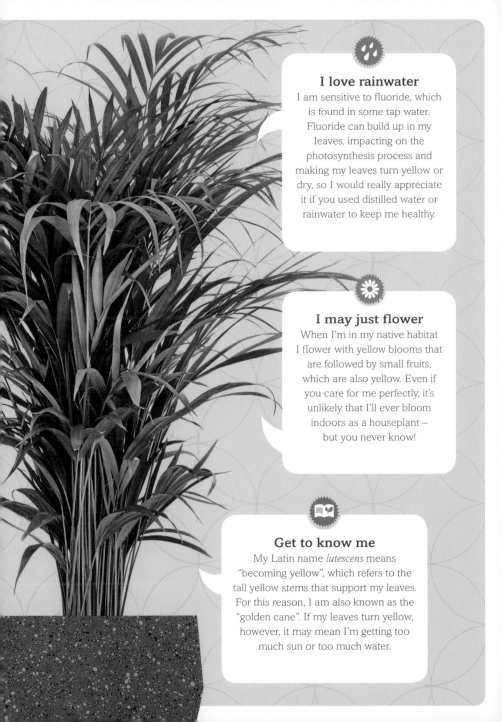

I love rainwater

I am sensitive to fluoride, which is found in some tap water. Fluoride can build up in my leaves, impacting on the photosynthesis process and making my leaves turn yellow or dry, so I would really appreciate it if you used distilled water or rainwater to keep me healthy.

I may just flower

When I'm in my native habitat I flower with yellow blooms that are followed by small fruits, which are also yellow. Even if you care for me perfectly, it's unlikely that I'll ever bloom indoors as a houseplant – but you never know!

Get to know me

My Latin name *lutescens* means "becoming yellow", which refers to the tall yellow stems that support my leaves. For this reason, I am also known as the "golden cane". If my leaves turn yellow, however, it may mean I'm getting too much sun or too much water.

POINSETTIA

Euphorbia pulcherrima

There are few houseplants as synonymous with Christmas as the striking red poinsettia. Found adorning many yuletide dining tables, near the Christmas tree, and on the front of hundreds of festive cards, this Mexican plant captures the essence of the festive spirit, with home and family at its heart.

I'll add colour

Although my recognizable red bracts are the most common and popular in the winter months, I can also be found in orange, yellow, pink, purple, and white. So whatever your décor, I'm sure you'll find a place for me in your room, where I'll add some star-shaped vibrancy whatever the colour.

❝I embody the **joyful spirit** of Christmas**❞**

Help me thrive

 POSITION
This plant needs to be grown in bright but filtered light. Do not let the temperature drop below 13°C (55°F), as this can damage the foliage.

 POTTING
Repot in early summer to the next pot size up, using a loam-based potting compost like John Innes No.3 with added bark or leaf mould.

 GROWTH RATE
These plants can grow to a height and width of 45cm (1½ft), growing reasonably fast in the right conditions.

 CARE
Keep dry after flowering. When in growth, water when the soil dries out, giving it a balanced liquid fertilizer feed every 10–14 days.

I bring good cheer

The sight of me will make you feel happy and festive as I am so closely associated with Christmas, family, and giving. My red colour evokes positive emotions of love and enjoyment and feelings of comfort and warmth, putting you in a relaxed and tranquil mood over the winter months.

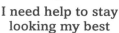

I need help to stay looking my best

It may be difficult to get my red bracts to colour for Christmas year after year, but to encourage me to perform my best, put me in a dark room with no artificial light after 12 hours of daylight every day for 10 weeks from November. And if you can't, I still make a pretty amazing green houseplant.

Be mindful of me

Cold blasts of air can hurt me, so please don't put me near a door or window that will be opened during the winter. When you're taking me home from the shop, lightly cover me so I don't catch a chill and freeze.

Get to know me

The Aztecs originally named me *cuetlaxochitl*, meaning "flower that withers", and also used the red pigment of my bracts to dye fabrics. Many mistake my brightly coloured leaves for flowers, but in fact they are bracts (modified leaves). I do have small yellow flowers, but they are not as beautiful as my bracts.

BE AWARE: I can be harmful if eaten, so keep me away from young children and pets.

WEEPING FIG

Ficus benjamina

The weeping fig is probably one of the most recognizable houseplants. The flexibility, beauty, and air-cleansing properties of this plant make it a sure-fire life-enhancer, while the delicate, glossy, oval leaves create a striking indoor feature.

I'll clean your air

My leaves can help reduce the risk of nose, mouth, and throat irritation caused by breathing in formaldehyde which comes from synthetic fabrics and xylene which can cause dizziness, headaches, and confusion, coming from paint, rubber, and printing.

> **"I'm a statuesque plant that's a great air purifier"**

Help me thrive

 POSITION
This plant needs a minimum temperature of 10°C (50°F) and to be placed where it will get indirect sunlight.

 POTTING
Grow in a loam-based potting compost like John Innes No.3, adding fine bark chippings to the mix to aid drainage.

 GROWTH RATE
Can reach 3m (10ft) tall by 2m (6½ft) wide, growing moderately fast, and can be pruned to remove dead material or to reshape and control.

 CARE
Don't overwater. Keep the soil moist when in growth, and water more sparingly in winter. Give it a high-nitrogen fertilizer every 4 weeks.

Be mindful of me

My leaves can fall if I suffer from a lack of food or light or if I'm left for a long time without water. When young I can be repotted every year to accommodate my growing roots, but when I'm mature I'll be happy staying in the same pot.

I have a positive energy

Me and my sister species of fig trees have been thought to possess positive power since ancient times and are important in many religions, including Islam, Hinduism, and particularly Buddhism, as Buddha was said to gain enlightenment under a fig tree.

Get to know me

I am from the mulberry family and have been around for a really long time – there are fossilized remains of *Ficus* plants dating back 60 million years. I am called a weeping fig because of my drooping habit. My foliage can fall like tears if I am stressed.

BE AWARE: My sap can be harmful, so keep me away from children, pets, and anyone allergic to latex.

RUBBER PLANT

Ficus elastica

Royalty of the houseplant world, this classic plant shines out from all others with its iconic large emerald-green leaves and tree-like habit. A rubber plant is certain to freshen up a tired spot, transforming it into a space of vibrant interest. It's one of the world's most popular indoor plants and deservedly so.

I'll remove toxins

My leaves remove carbon dioxide and may also help remove airborne bacteria and mould spores. I can also help remove formaldehyde and benzene from the air, which are emitted by paint, cosmetics, glues, and detergents and can cause headaches, drowsiness, and irritate eye, nose, and throat.

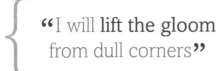

" I will **lift the gloom from dull corners "**

Help me thrive

 POSITION
Grows best in light shade, at 15–24°C (60–75°F); do not let the temperature dip below 13°C (55°F), but keep away from radiators.

 POTTING
Repot annually during the first 5 years, then every 2 years, into a container one or two sizes bigger and using well-draining potting compost.

 **GROWTH RATE**
Can eventually grow to ceiling height, so it may need a cane to support it or a wall to lean on. Prune in late winter to keep it in check.

 CARE
Feed every 2 weeks from spring to early autumn with a diluted liquid fertilizer. Water little and often, and never let it sit in water.

Be mindful of me

I use the dark green pigment in my leaves to absorb the light that gives me energy to thrive, so please keep my leaves free of dust by wiping my foliage clean every month so I can stay healthy.

Give me a trim

When I reach the size that perfectly fits my home, you can keep me at that size by cutting off my top and pruning any branches to make me look fuller and healthier. I'd appreciate it if you did this in spring or summer.

Split me in two!

When I get too tall, you can split me into two plants by "air layering" me. Remove two leaves where you want to halve me. Strip away a 1cm (½in) ring of bark, then pack a handful of damp moss around me and cover with a plastic bag. When roots appear, cut me below the roots and pot up my top section as a new plant.

Get to know me

If you cut into my bark, I will drip a substance called latex, which can be used to make rubber. But avoid touching this sap, as it's an irritant.

BE AWARE: My sap is harmful if eaten, so keep me away from children and pets.

CAPE JASMINE

Gardenia jasminoides

There is no doubt that this is one of the most beautifully fragrant indoor plants. The matt white flowers are so perfectly formed and contrast strikingly with the sea of glossy, emerald-green leaves they are nestled in. Still, their beauty cannot compare to the spirit-raising power of their scent.

I'll relax you

My scent is renowned for inducing relaxation, which is why gardenia essential oils are frequently used in the cosmetics industry in lotions, perfumes, and body washes. My fragrance fills your nose and heart with beauty and will soothe you as you unwind. It has even been extracted to treat depression and anxiety.

> **" Feel uplifted by my stunning fragrance "**

Help me thrive

 POSITION
Grows well in cool, bright rooms. To flower, it needs a minimum day temperature of 21°C (70°F) and at least 16°C (60°F) at night.

 POTTING
Repot into the next size up in spring when the roots fill the pot, as they flower better with congested roots. Use an ericaceous compost.

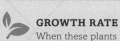 **GROWTH RATE**
When these plants are kept indoors, they can reach 2m (6½ft) tall and 1m (3¼ft) wide. They are moderately fast growers.

 CARE
Never let plants dry out or get waterlogged. Water them freely with rainwater when in growth, applying a balanced liquid fertilizer every 4 weeks.

Help me feel at home

I originated in Asia and grow well in subtropical climates. I like high humidity but I don't like sitting in water, so please sit me on a saucer of wet gravel or pebbles; the water will evaporate with the heat and create humidity in the air around me.

I'm a fashion statement

I'm a chic, sophisticated addition to any room and I'm loved by trendsetters, including fashion design icon Coco Chanel and legendary jazz singer Billie Holiday, who decorated her hair with my flowers to create her trademark style.

Help me to re-flower

Once I've stopped flowering, cut off any dead flowers to encourage new blooms and prune any leggy old shoots, leaving my new growth to keep me looking healthy and compact. And don't wait too long to prune me, because it might mean I don't flower next year.

I symbolize love

Thanks to my white flowers and delicious scent, I am believed to symbolize love, purity, and sweetness, which is why I am often used in table centrepieces or bouquets at weddings.

BE AWARE: I can be harmful to pets if eaten, so keep me away from cats and dogs.

GUZMANIA

Guzmania cultivars

I have used this striking South American plant in many interior design landscapes that I've created over the years, because it gives so much positive energy in its structure and vibrant colours. The other main benefit is that the rosette-shaped, flower-like bracts, or modified leaves, stay bright red for such a long time with minimal upkeep.

> **"** Try me if you're looking for a **pop of colour "**

I'll clean your air

My leaves remove air pollutants such as benzene, which is found in glue and paint, and release oxygen at night. This makes me ideal for the bedroom as I will purify the air while you sleep, helping you to wake feeling refreshed.

My flowers are hard to spot as they don't rise above my bracts

Help me thrive

 POSITION
Grows well in high humidity and in indirect sunlight – direct sunlight will burn the leaves. Keep temperatures above 15°C (60°F).

 POTTING
Likes a rich, free-draining compost like John Innes No.1 mixed with equal amounts of chopped sphagnum moss.

 GROWTH RATE
Some varieties can reach 60cm (2ft) tall and 1.2m (4ft) wide; they are fairly slow-growing and long-lived, but once they've flowered, they will die.

 CARE
Quarter-fill the central urn with rain or distilled water in summer and refresh monthly. In winter, water sparingly to keep soil moist.

Let's make babies

I am monocarpic, which means I die after my true flowers have bloomed. But I can be easily propagated from the offsets that grow at my base. When the pups get to 10–15cm (4–6in) they can be cut away and planted into a free-draining potting mix.

Get to know me

As a bromeliad, I am an epiphyte, which means I get my nutrients from the air and the rain. So I have a funnel-shaped urn formed by the centre cluster of my foliage. In my natural habitat, this is where I collect rainwater to drink. In Brazil, my urn is believed to catch all of the "blessings from above".

I will brighten your day

My star-shaped bracts will add intriguing structure to your interior, and my vibrant colours – I'm available in a huge variety, such as red, orange, purple, green, pink, white, or even bi-coloured – will energize you.

Look out for my true flowers

My brightly coloured bracts may be my main attraction, but I do also produce true flowers. They are small and white and bloom in between my bracts, and when they appear, this signifies the approaching end of my life, which will take about a year.

VELVET PLANT

Gynura aurantiaca

This is the closest you can get to a plant that looks as if it's emitting ultraviolet light. The hairs on the leaves are so vivid they appear to glow – it's almost supernatural! The abundance of foliage makes it a real talking point, creating a display that turns a dull windowsill into a strong point of interest.

{ **"I will be a source of constant wonder"** }

Help me thrive

 POSITION
Place where it'll get good indirect light for a few hours a day, especially during winter. Needs a temperature of at least 13°C (55°F).

 POTTING
Can be potted on in spring into a loam-based compost like John Innes No.2, or replaced by propagated offsets if it grows too leggy.

GROWTH RATE
This is fast growing so cut back or pinch out growing tips if it becomes too leggy. Given support, it can reach 2m (6½ft) high and 1.2m (4ft) wide.

CARE
Water freely during the growing season to keep the soil moist. Its roots are sensitive to rot, so never overwater or allow to sit in water.

I'll add some magic
A rare colour in nature, my almost magically purple leaves evoke feelings of creativity, fascination, and wonder. Keep up the bright, indirect light, otherwise my hue may fade, and try not to get my leaves wet as they scorch easily.

I can get hungry
I am an incredibly hungry plant during spring and summer. Please give me liquid fertilizer every one to two weeks during this growing season, to encourage me to produce more of my extraordinary foliage.

Touch me!
My common name is the velvet plant because I have many vibrant purple little hairs on my leaves, which give me a soft, velvety feel. Interact with nature by touching my soft leaves for even more sensory magic.

*My pungent,
thistle-like flowers
bloom in winter*

My flowers are unusual

My flowers are as fascinating as my foliage. They look like thistles and are vivid orange, forming a great contrast to my leaves. Their fragrance, however, is another matter – some people choose to snip off my blooms because they're not too keen on the powerful scent.

Give me a new lease of life

I lose my vibrancy and turn darker as I mature. Keep me looking good as new by propagating me. Cut a stem at around 6–10cm (2½in) long and place it in water, then pot it up once roots have formed.

AMARYLLIS

Hippeastrum cultivars

Everyone should grow at least one amaryllis in their lifetime – they are truly dynamic. Growing this indoor bulb instils a real sense of pride because of the speed of its growth and the magnificence of its flowers. There are so many different varieties and colours to choose from – they are a plant collector's dream.

I'll make you proud

I am a marvellous plant for promoting mental wellbeing as I am easy to grow and unlikely to fail, giving you a great sense of pride and achievement. You won't have to wait long for your reward, as my spectacular blooms are quick to appear.

{ **"Take pride** in cultivating my magnificent **blooms"** }

Keep me going

After I have flowered in winter or spring, please remove the flower stalk to within 2cm (½in) of the top of the bulb. Begin feeding me every two weeks with liquid fertilizer once I start shooting new growth. This will help me refill my bulb with energy.

Help me thrive

 POSITION
Will grow well on a windowsill in bright filtered or full light and needs a temperature of at least 12°C (54°F).

 POTTING
Pot on in autumn into a pot 5cm (2in) wider than the bulb, with the bulb half out of the compost – but only every 3–5 years to minimize root disturbance.

I don't like it too wet

Like all bulbs, if I sit in soggy soil it can lead to rotting, so please don't leave me in a pot cover or saucer with a pool of water in it. During my dormant period when my leaves are cut, you can leave me to rest with no water at all.

 GROWTH RATE
Can reach 30–60cm (1–2ft) tall and 30cm (1ft) wide. They tend to bloom in winter or spring, 6–10 weeks after the bulb has been planted.

 CARE
Water when the top of the soil feels dry and feed every 10 days when leaves start to appear. Stop watering in early autumn and remove old leaves.

I'm a showstopper

My name amaryllis comes from the Greek, meaning "sparkling", for my breathtaking multi-headed blooms, which can be found in red, orange, and pink. You'll be amazed that a bulb can produce such beauty.

Let's be strong together

I am commonly thought to signify determination and strength, making my flower the perfect symbol for the Huntington's disease community across the world. I represent hope and help to raise awareness of advancing research on the condition.

BE AWARE: My bulb is harmful if eaten, so keep me away from children and pets.

PLANTS TO CHEER YOU UP

Some plants can change our mood and brighten our day in an instant. It may be the sweet fragrance of a bloom or the cheeky shape of foliage. It could even be the simple joy of seeing your favourite variety flower. These delights can lift your heart and make you smile.

Amaryllis ▶
Hippeastrum cultivars

With so much energy and life within its bulb, it is hard not to be cheered up by this plant. Easy to grow, a joy to watch, and bringing some much-needed colour in the winter, it soon becomes part of the family. *See pp.92–93*

Cape jasmine ▲
Gardenia jasminoides

The fresh, sweet fragrance of the Cape jasmine is sure to cheer up anyone who steps through the door. One sniff and bang! You're on holiday in Greece, walking down to the *taverna*. The heavenly scent of this plant really is aromatherapy for the soul – it will delight your senses and fill you with joy instantly. Place it where you and your visitors will frequently walk past it for maximum impact. *See pp.86–87*

Living stones ▼
Lithops spp.

This plant will entertain you and your guests. They are funny to describe and even funnier to see. Whether you think they look like brains or bottoms, they are sure to brighten your day and make you crack a smile – especially when you see the flower blooming from between the two "stones". *See pp.102–103*

Hearts on a string ▶
Ceropegia linearis
subsp. *woodii*

Whether on a shelf or in a hanging basket, the heart-shaped leaves of this plant trailing over their pot will fill you with joy and affection, reminding you of the loved ones in your life. Its unique and unusually shaped flowers will make you laugh, bringing happiness to your home.
See pp.52–53

The sensitive plant will fascinate you

Sensitive plant ▲
Mimosa pudica

The wonder of nature is certainly a reason to feel joy, and the remarkable interactivity of the sensitive plant as it reacts to you and the world around it really makes you feel connected with nature. The way the leaves move and curl in response to your touch will surprise and fascinate you – even if you've seen it hundreds of times before!
See pp.106–107

POLKA DOT PLANT

Hypoestes phyllostachya

This is a perfect small foliage plant. It's colourful, dainty, and flexible in its uses too. A lively, uplifting little plant for a windowsill or table, it's also fantastic planted with other indoor varieties in baskets, arrangements, bottle gardens, or terrariums as a fascinating feature.

> **"I'm cheering and versatile – a patterned wonder"**

I am fun and cheerful

My playful, patterned pink leaves will add a softness to your room, while also evoking feelings of child-like joy and happiness. The spotty patterns on each of my leaves are completely unique and individual, just like you are!

Keep me colourful

To keep my foliage vibrant, place me in partial rather than full sun. This may make me leggy, however, as my stems seek out the sun: trim them to make me fuller.

Help me thrive

 POSITION
Indirect light in humid conditions is best, so mist daily from spring to early autumn and occasionally in winter, ideally with rainwater.

 POTTING
Repot when rootbound – you may spot this if it stops growing in summer. Pot into the next size up with a free-draining compost.

 **GROWTH RATE**
Plants will reach about 45cm (1½ft) high and 30cm (1ft) wide. Pinch out the growing tips frequently as this will encourage a bushier habit.

 CARE
Water to keep the soil moist when growing and sparingly in winter. Give a liquid feed monthly in summer, and prune in spring to boost regrowth.

If left to grow, my tubular flowers appear on spikes

My artistic leaves will inspire you

I am also commonly known as "freckle face", due to my bold, patterned leaves, which often feature red, purple, and pink speckles. My spots commonly blend together to merge into big areas of colour – just like a spray painting.

Look out for my flowers

I may surprise you with my flowers. They bloom from spikes and are lavender-coloured, and can mark the moment when I start approaching the end of my life – snip them off to preserve my energy.

Let's make babies

I am easy to propagate by sowing the seeds from my flowers, or you can take cuttings from my branches and pot them into a cactus mix, where they will readily root.

MEXICAN HAT PLANT

Kalanchoe daigremontiana

This amazing plant has large, fleshy leaves, grows quickly, and is easy to keep. It is known as "mother of thousands" – and you can see why, as it produces hundreds of baby plants around its leaf edge. Grow these on to share the love and connect with friends and family.

> **❝** I am **one** plant that will give birth to **many** **❞**

Help me feel at home

I am native to Madagascar, so I am used to growing in rocky and dry places. That's why it's best to pot me up using a cactus potting mix. Alternatively, you can try adding sand to an all-purpose mix to help drainage.

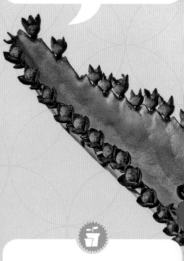

Help me thrive

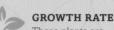

 POSITION
A sunny windowsill is ideal for this plant. It can even spend the summer outside, but make sure it's gradually introduced outdoors.

 POTTING
Repot into the next size up when the roots outgrow the container, using a loam-based compost with added grit.

 GROWTH RATE
These plants are fast growing and can ultimately reach a height of 1m (3¼ft) and a width of 30cm (1ft).

CARE
Water when the soil starts to dry out in summer and feed monthly with a balanced liquid fertilizer. Reduce watering to a minimum in winter.

Let's make babies

I only live for around two years, but I keep going by producing more baby plants than you can handle. Plantlets will keep falling off me, so fill pots or a seed tray with cactus compost and lightly touch the soil with the base of a plantlet – it will soon take.

I'm out of this world

I was sent into space to *Salyut 6*, the Soviet space station, in 1979 to boost the mood and morale of the crew. Originally sent for research purposes, plants like me were soon found to help calm the cosmonauts on board and alleviate loneliness and depression.

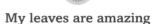

My leaves are amazing

I rarely bloom when kept indoors – it's my water-storing succulent leaves that I'm known for. Their edges are fascinating – they look almost like a spinal cord with lots of segments, which explains my other common name: "devil's backbone".

I like a holiday

Although I'm a happy houseplant, I'd quite like to be outside for the summer. Gradually introduce me to the outdoors by first letting me sit in a little morning sun, then place me in bright, indirect light. Bring me back inside before temperatures drop below 4°C (39°F).

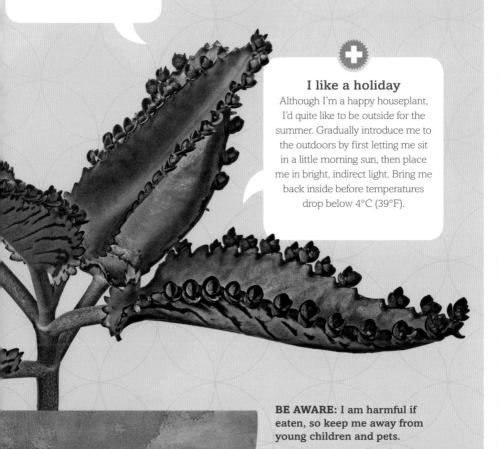

BE AWARE: I am harmful if eaten, so keep me away from young children and pets.

FLAMING KATY

Kalanchoe blossfeldiana

This is a special little plant with fleshy, shiny, deep-green leaves. The flowers grow in clusters and are full of colour, available in white, red, yellow, orange, pink, and more. They are uplifting to look at both from a distance due to their bright colour and close up because of the enchanting detail of the blooms.

I'm a great gift

I am a cheerful little plant. I brighten up every room with my different-coloured blooms. And because I also symbolize wealth, prosperity, and lasting affection, I make a very popular gift for every occasion, as a sweet token of love.

> **"I am a bright, cheerful gift for all occasions"**

Help me thrive

 POSITION
Grows well on a sunny windowsill out of direct sunlight and is happy with the humidity and temperature of an average room.

 POTTING
Repot into a container the next size up every year using a loam-based compost such as John Innes No.2. Add grit to help with drainage.

 **GROWTH RATE**
This plant is fast growing and can reach a height and width of 40cm (1¼ft). Prune back after flowering to keep foliage bushy and dense.

 CARE
When in growth, water regularly when the soil has just started to dry out and feed weekly with a balanced liquid fertilizer. Water sparingly in winter.

I'll add colour to your life

I'll never cease to surprise you and put a smile on your face as I flower repeatedly. My blooms are so very bright and colourful, and their brilliance is only enhanced by the contrast with my waxy emerald-green leaves.

Keep me flowering

To help me reflower, cut off my spent flowers as soon as they have died to encourage new shoots. It's possible to keep me in flower all year if you can give me 12 hours of total darkness daily in late spring and summer, as I flower in the wild when the days are shorter.

Get to know me

As a succulent, I can tolerate being a little dry because I store water in my leaves. This makes me very easy to look after, so I'm a great plant for beginners or teenagers.

Let's make babies

You can propagate me easily by taking leaf cuttings and potting them straight up or rooting them in water first, which takes around two weeks. To ensure good drainage, use a cactus potting mix or an all-purpose compost with added sand.

BE AWARE: I am harmful if eaten, so keep me away from children and pets.

LIVING STONES

Lithops spp.

This is a truly weird and wonderful plant, full of surprises and sensory stimulation. The soft, padded leaves are covered with mottled green or reddish tones, with a margin of grey or brown. As a kid, I thought it looked more like little brains than the "living stones" of its name!

I'm one of a kind
I don't look like any other houseplant. I am such an interesting shape and colour, I'll inspire a real sense of wonder at the variety of Mother Nature, taking your mind away from the stresses and pressures of everyday life.

> **"** My **bizarre** appearance will **amaze** and delight **"**

Help me thrive

 POSITION
Needs 5–6 hours of direct sunlight a day, so place in a sunny spot on a south-facing windowsill. Also prefers low humidity so keep somewhere dry.

 POTTING
Lithops hardly ever needs potting on, but if the plant becomes crowded, wait until late spring and use a cactus compost mixed with grit.

 GROWTH RATE
These are very, very slow-growing plants, reaching no more than 3cm (1¼in) tall by 3cm (1¼in) wide at their maximum size.

 CARE
Do not water at all from early autumn to mid-spring. The rest of the year, soak it every 2 weeks, letting it fully dry out between waterings.

Get to know me
My Latin name is derived from the Ancient Greek term *lithos*, which means "stone", and *ops*, which means "face", referring to my stone-like appearance. You can buy me as a small plant or grow me from seeds, which form when my flower dies back and the seed pod ripens.

Touch me

I am a tactile plant. I look like a soft pebble and when you touch me you can feel the small dimples and indentations on my skin. As a succulent, I'll feel firmer when I'm hydrated; if I start to wrinkle and soften, it may be because I need a drink.

Look out for my flowers

Surprisingly, I produce delicate, daisy-like flowers with yellow centres. Once I reach three years old, I will flower in summer or autumn. My blooms usually have a sweet scent, and once they fade I become dormant until two new leaves grow and the cycle of life repeats.

My daisy-like flowers burst from the fissure between my leaves

Don't overwater me

I store water in my leaves, so be careful not to overwater me. From late spring to summer, do water me but leave my soil to dry out before doing so again. And please water me in the morning so there's time for any excess water to evaporate.

PRAYER PLANT

Maranta leuconeura
'Fascinator Tricolor'

At 16, I got a job in a garden centre and bought this plant with my first pay packet. I was captivated by its delightful foliage, whose sweeping scarlet veins looked hand-painted. And when the leaves closed up each night, it felt like having a true living companion.

I will fascinate you

I have the heartwarming ability to fold my leaves upwards in the evening as if I am praying. I can do this because of a hinge-like structure at the base of my leaves, which is triggered at night when my leaves drain of water. You may hear me make a rustling sound at the same time.

{ **"**My **folding** patterned leaves will **enchant** you**"** }

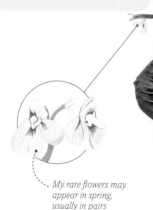

My rare flowers may appear in spring, usually in pairs

Help me thrive

 POSITION
Place out of direct sunlight and draughts. Ideally keep in a high-humidity room; otherwise, mist leaves regularly in summer.

 POTTING
Pot on to the next size up when the roots grow through the bottom of the container or in circles inside. Use a standard potting compost.

 GROWTH RATE
Can reach a height and width of 30cm (1ft). Give leaves an occasional prune just above a leaf node to encourage bushier regrowth.

 CARE
Feed monthly with a balanced liquid fertilizer during the summer and water moderately, to keep the soil moist. Water sparingly during winter.

I'm a work of art

My elegantly patterned leaves are also often red underneath – you can catch a glimpse of this when my leaves grow horizontally and fold upwards at night.

Help me feel at home

As a native of the tropical forests of Brazil, I love indirect light and humidity. My colour intensity is enhanced if I am not in full sun. You can also replicate my forest home by grouping me with other plants or placing me in a tray of wet pebbles to boost humidity levels.

Let's make babies

Propagating me is easy. Simply make a stem cutting below my leaf and place it in a jar filled with water until it roots. My baby can then be planted in fresh potting soil, where it will grow fairly quickly.

My flowers are a rare treat

I can grow dainty pink-white tubular flowers, borne on stems to elevate them above my foliage, although they are very rare when I'm grown as a houseplant.

SENSITIVE PLANT

Mimosa pudica

When I first saw this plant move, I was gobsmacked. A plant that will fold its leaves inwards when you touch it? How unplantlike, as though it's from another world! It's a constant reminder of the wonders of nature and a calming, characterful presence in your home.

> **"I am a magical plant; I respond to your touch"**

I'm alive

I'm a unique plant that will be a living presence in your room, because I can open and close my leaves when touched or shaken or there's a change in temperature. This gives me a human or animal-like quality, so you are bound to emotionally connect with me.

Help me thrive

 POSITION
Keep out of direct sunlight and at temperatures no lower than 13°C (55°F), ideally in a humid location, such as a bathroom.

 POTTING
Repot into the next size up when roots grow through the bottom of the pot or in circles inside. Use a soil-based compost like John Innes No.2.

 GROWTH RATE
Can reach 60cm (2ft) tall by 60cm (2ft) wide. Taller growing stems might need cane support or cutting out.

 CARE
Keep the plant moist at all times and feed monthly during the summer with a general liquid houseplant fertilizer.

Let's make babies

I can easily be propagated by taking semi-ripe cuttings, which are from the current season's growth, in summer or autumn. Plant these cuttings in seed and cutting compost, which will help them to establish strong roots.

I'll relax you

I have a low, spreading habit and my dainty fern-like leaves will add some soft, serene texture to your room to help you relax. My vibrant, deep-green leaves will also rejuvenate and revitalize you when you're feeling exhausted.

Be mindful of me

Although it's fun and inspiring to watch me shrink away when touched, it can actually weaken me, as it takes up my energy to open and close, so try not to do it too often and feed me so I can keep growing.

Look out for my flowers

In my native Brazil, I bloom tiny pink or pale-purple flowers like fluffy balls, though I may not produce any as a houseplant.

BE AWARE: I can be harmful if eaten, so keep me away from children and pets.

SWISS CHEESE PLANT

Monstera deliciosa

This marvellous plant will bring the drama and cleansing power of the South American rainforest into your home. The name *Monstera* means "monstrous", referring to the huge size it can grow to, while the holes in its leaves have earned it the fun name Swiss cheese plant.

{ **"My vast leaves** bring the **rainforest** indoors **"** }

Help me thrive

 POSITION
Place in a moderately bright spot with filtered, indirect light and space to grow; leaves can turn yellow in too much light.

 POTTING
When young, repot annually; when mature, repot when rootbound, using a rich compost such as John Innes No.3 mixed with fine bark chippings.

 GROWTH RATE
Can reach heights of 3m (10ft) and a width of 75cm (2½ft), growing around 30cm (1ft) every year. Taller growing stems may need cane support.

 CARE
Water when the soil starts to dry in spring and summer, and feed monthly with a general liquid fertilizer. Reduce watering in winter.

Keep me clean
My glossy foliage is great at trapping any dust floating around your room. You can then clean my leaves lightly with a damp cloth to keep them looking vibrant, so that you and I can both breathe more easily.

Let's make babies
You can propagate me by cutting one of my mature stems below an aerial root (which are roots that grow above the soil). Do this during summer and plant the cutting in moist but well-drained houseplant potting mix, where it will root and grow happily.

I'll clean your air
I have huge leaves, which means there is a large surface area that can take in carbon dioxide and release oxygen into the atmosphere, making me highly efficient at purifying the air in your home.

Get to know me

I am a semi-epiphyte, so I can live both independently and on the surface of other plants. This means I can use my aerial roots to climb or even be trained. If I don't get enough light my new leaves will grow towards the dark rather than the light. This is known as negative phototropism, and it's how I find trees to climb up in the rainforest, as I grow towards their shade.

BE AWARE: I am harmful if eaten, so keep me away from children and pets.

SWORD FERN

Nephrolepis exaltata

The sword fern or Boston fern is one of the iconic Victorian parlour plants, with its lush, refreshing green fronds that arch so gracefully. This humidity-loving plant has multiple benefits, softening corners, adding impact to dull spaces, and drastically improving air quality. It is a truly stunning fern.

I'll remove toxins

I am very effective at preventing "sick building syndrome" – which is when people develop symptoms such as headaches and respiratory problems caused by their building environment. I can neutralize toxins in the air such as formaldehyde and xylene, which are often found in building materials and new homes.

{ **"**My leaves bring **serenity** and **clean** your air**"** }

Help me thrive

 POSITION
Needs filtered light and grows well in warm, humid rooms, so keep it above 10°C (50°F) in winter, but below 25°C (77°F) in summer.

 POTTING
Repot into the next size up when roots appear through the bottom of the existing pot or grow in circles inside. Use a rich but free-draining compost.

 **GROWTH RATE**
This fast-growing fern can reach a height and width of 60cm (2ft). Trim off old, discoloured fronds from around the base to allow new growth.

 CARE
Keep the soil moist with rainwater or distilled water. If it drops to 15°C (60°F) or below, let the top third of the soil dry out before watering again.

Let's make babies

Look out for my baby plants or runners around the edges of my pot. Let them grow until they are around 5cm (2in) long and strong enough to cut off, then you can pot them up in a container using a houseplant potting mix.

Keep me beautiful

Keep my fronds clean so they are better able to take in light, but avoid chemical wipes as I am quite sensitive; give me a quick shower instead. And turn me frequently, so every part gets light, to encourage symmetrical growth.

I will calm and soothe

Plant me in a hanging basket or place me on a pedestal so that my fronds can trail over the sides and cascade down. This will give a beautiful weeping-willow effect, providing a gentle, soothing presence in the room.

Help me feel at home

I like warmth and humidity, and happily grow in a bathroom or kitchen. In other rooms, you can mist me regularly or place me near a humidifier.

MOTH ORCHID

Phalaenopsis

This is my favourite type of orchid. In Greek, *Phalaenopsis* means "moth-like", which is apt, because this plant's flowers look like moths in flight. Flowering spikes grow from the tongue-like foliage, bearing many flowerbuds that open in stages right to the tip. An easy plant to keep, it will clean your air and, if fed and watered, flower for months.

I'll remove toxins

I can help maintain healthy breathing functions by reducing formaldehyde released from composite wood products, as well as xylene and toluene, which are used in paints, paint thinners, and varnishes. Scientific studies have shown me to be one of the best air-purifying houseplants.

{ **"** I'm a **champion** bloomer and expert **air-purifier "** }

Be mindful of me

Keep my leaves clear of dust by wiping them with a damp cloth – this will allow maximum light to reach my foliage as well as keep me looking my best. As I am native to South-east Asia and the Philippines, I also like humidity, so mist me regularly with rainwater, please.

Help me thrive

 POSITION
Likes bright, filtered light and a warm centrally heated room of at least 19°C (66°F). Move to a shadier spot during summer.

 POTTING
Repot in spring into the next size up in a free-draining orchid mix when it has outgrown its pot or been in the same soil for 2 or more years.

GROWTH RATE
Plants can grow to anything from 15cm (6in) to 1m (3¼ft) tall and to a width of 20–30cm (8–12in), at varying speeds.

CARE
Water only once the soil begins to dry out and feed every 2 weeks with a liquid orchid fertilizer when in growth. Reduce both in winter.

Get to know me

As an epiphyte, I get my moisture and nutrients from the air, so you will often see my aerial roots growing above the soil as well as below. Please don't cut off or bury these wandering roots! I am often sold in a clear pot, which I prefer as my roots are attracted to light and help perform photosynthesis.

My flowers are long-lasting

I can flower at any time of the year, sometimes up to three times, and my blooms can stick around for three months or more. Once I have finished flowering, cut back the stem to the biggest bud and I might give you another flush of flowers.

Let's make babies

Sometimes I will grow little babies at my base or along my stem, which will develop and grow their own roots. These are called *keikis*, and once they have several roots you can remove them and pot them up to flourish in their own containers.

HEART-LEAF PHILODENDRON

Philodendron cordatum

This flexible plant will do well in most households. You can either display it as an indoor climbing plant or leave it cascading for an elegant and calming effect. Its immensely pretty heart-shaped leaves are also air-purifying.

I'll remove toxins
My leaves can help alleviate asthma symptoms by removing formaldehyde from the air and releasing oxygen. Formaldehyde is a toxin that can permeate the atmosphere of your home, especially from processed wood furnishings.

{ **"**I am **versatile** – I can both climb and trail**"** }

Help me thrive

POSITION
Give plenty of growing space in a part-shaded position out of direct sunlight and don't let temperatures drop below 15°C (60°F).

POTTING
Repot into the next size up when roots grow out of the bottom of the existing container, ideally in early summer. Use general potting compost.

GROWTH RATE
If given good growing conditions, this plant can grow easily and quickly, reaching 3m (10ft) tall (or long) indoors.

CARE
Keep the soil moist when in growth using lukewarm water and feed monthly with a balanced liquid fertilizer. Water sparingly in winter.

Be mindful of me
My leaves turn lighter green and yellow if I'm hungry. I also like a spritz and a wipe occasionally to keep my leaves dust-free.

You can train me

If you would like to keep me as a tall plant, you can train me to grow onto a moss pole. However, if you would prefer me to trail, either plant me in a hanging basket or position me on top of a tall cupboard or table so I can cascade down.

Let's make babies

I am easily propagated by taking stem cuttings with a few nodules on them. Place these in water until roots appear from the nodules. Then the cuttings can be planted up in a well-drained potting mix.

I like a holiday

I will enjoy a trip to the garden during the summer months, though sit me in a shaded spot. Unlike many houseplants, I can tolerate the change between indoors and outdoors relatively stress-free.

BE AWARE: I am harmful if eaten, so keep me away from young children and pets.

BLUE TORCH CACTUS

Pilosocereus azureus

Native to Brazil, this beautiful cactus has upright stems of turquoise embellished by translucent golden needles. Although it gets big in its natural habitat, its slow-growing nature is ideal for indoor pots and it will be your cool, calming, steadfast friend for many years.

I'll be an enduring friend
Age and resilience are my best assets. Because I am easy to keep and grow slowly, I am likely to enjoy a long life and will be with you for years, becoming a well-loved member of the family – just be careful of my spikes and keep me away from young chidren.

Make a feature of me
Place me somewhere prominent and, with my blue hue and candle-like appearance, I will prove an intriguing talking point for visitors.

{ **"I'm soothingly serene and long-lasting"** }

Help me thrive

POSITION
Will grow best in full sunlight in warm or hot temperatures of 20°C (70°F) or higher and preferably not below 15°C (60°F).

POTTING
Use a standard free-draining cactus mix. Cacti always look better in clay pots – they are heavier and help prevent them from toppling over.

GROWTH RATE
Vigorous and fast growing, it will quickly reach a height of 30cm (12in), but in the right conditions it can reach 1m (3ft) or more.

CARE
Water weekly in spring and summer with rainwater or distilled water, but don't overwater as it may cause rotting. Keep drier in winter.

My blooms develop into fleshy fruits

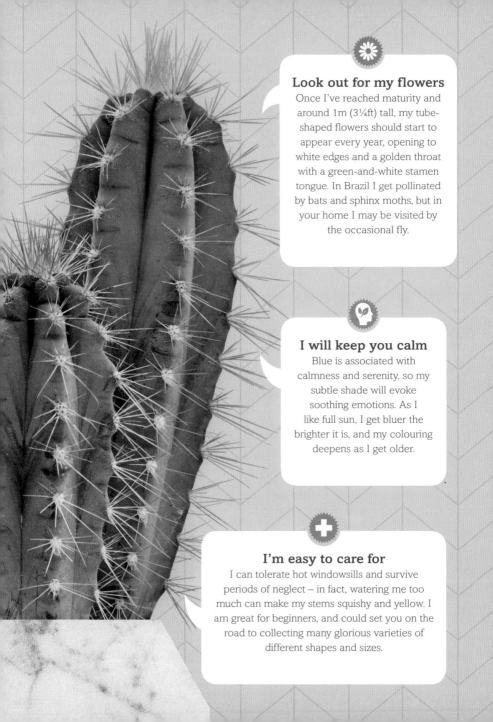

Look out for my flowers

Once I've reached maturity and around 1m (3¼ft) tall, my tube-shaped flowers should start to appear every year, opening to white edges and a golden throat with a green-and-white stamen tongue. In Brazil I get pollinated by bats and sphinx moths, but in your home I may be visited by the occasional fly.

I will keep you calm

Blue is associated with calmness and serenity, so my subtle shade will evoke soothing emotions. As I like full sun, I get bluer the brighter it is, and my colouring deepens as I get older.

I'm easy to care for

I can tolerate hot windowsills and survive periods of neglect – in fact, watering me too much can make my stems squishy and yellow. I am great for beginners, and could set you on the road to collecting many glorious varieties of different shapes and sizes.

AFRICAN VIOLET

Saintpaulia hybrids

My gran grew these cute little plants – her windowsills were full of them – so I always knew what gift to buy for her. It's such an easy plant to care for, and will consistently reward you with vibrant flowers. With so many different types and colours to collect, there is always something new to make you smile.

I will cheer you up

Although I can flower at any time of year, it's usually between mid-autumn and early spring, bringing some colour and warmth to the colder months. I am commonly found in dark purple tones, which, as a mix of red and blue, strikes the perfect balance between energizing and relaxing.

{ **"**I am a **cute** little plant bursting with **flowers"** }

Help me thrive

POSITION
Thrives in warm, humid conditions, above 15°C (60°F), and out of direct sunlight in summer. Give it the brightest light available in winter.

POTTING
Repot at least once a year but use virtually the same container size, as it flowers better when the roots are tightly packed.

GROWTH RATE
Most hybrids are relatively small and slow growing, reaching heights and widths of around just 15cm (6in).

CARE
Allow soil to dry out before watering. Feed fortnightly with a liquid fertilizer high in potassium and phosphate. Water occasionally in winter.

Get to know me

I'm not actually a violet, it's just that my flowers have a striking resemblance to true violets. As my name suggests, I was first found in the hills of Tanzania in East Africa, but in the wild I don't just grow in tropical places. I can also grow in the mountains, sheltered under other plants.

My flowers steal the show

I can also be found in pink, white, blue, and red, and even with different patterns on my flowers, such as stripes, or with my petal edges tinted. Look out for my double-flowering varieties, too. To encourage me to bloom, I should get about 10 hours of indirect light a day.

I'm fussy about water

Firstly, I like lukewarm water, as very cold water can chill me. I can also rot if I have too much, so water me from the bottom and never let me sit in water. Allow my soil to dry out between waterings to encourage me to flower, and please don't get water on my hairy foliage, as it can mark.

Let's make babies

Make new plants from me by taking leaf cuttings with the stalk attached. Cut off the top of the leaf to encourage faster root growth, then pot it up in seed or cutting compost. More than one plantlet may grow, and once they are big enough to handle, repot them into an all-purpose potting mix.

SNAKE PLANT

Sansevieria trifasciata

This is probably the toughest houseplant you can buy – it's low maintenance, tolerant of neglect, and a hard-working air-purifier. It also has an interesting form, with its upright sword-like foliage, and fascinating variegated leaf markings that will provide a positive mental distraction from your worries.

I'll remove toxins

I can help reduce the damaging health effects of trichloroethylene, a chemical compound that is found in some household degreasing cleaning products, which can cause nausea with high levels of exposure. I also remove carbon dioxide and produce oxygen throughout the night.

{ **"**I'm a **low-maintenance** delight for beginners**"** }

Help me thrive

 POSITION
This plant will grow well in a sunny position but will also tolerate some shade. It can handle the dry air in our homes as well as more humid spots.

 POTTING
Repot into the next size up when roots are growing through the bottom of the pot or circling inside. Use a free-draining, gritty compost.

 GROWTH RATE
This is slow growing but will grow faster if its compost is replaced regularly. It can reach up to 1.2m (4ft) and spread to 50cm (1½ft).

 CARE
Let the soil dry out between waterings. Feed once in spring and once in summer with a liquid houseplant fertilizer. Keep leaves free of dust.

Let's make babies

You can propagate me by snipping off a healthy leaf that's about 5cm (2in) long and inserting the base of the cutting into a free-draining gritty compost or just water. You can also grow new plants from division, where you take me out of my pot and split me up.

I can cope with neglect

I am extremely low-maintenance and will tolerate a bit of neglect. In fact, the only thing I don't like is being overwatered, so always let my soil dry out before watering me again. You may be able to spot if I've been overwatered as my leaves will start to go soggy or mushy.

I'll protect you

In feng shui, I am believed to have a strong protective energy. The spiky nature of my leaves is said to shield you from negative *chi*, although this is sometimes interpreted as an aggressive energy, with some people suggesting I shouldn't be placed in busy social areas.

BE AWARE: I can cause nausea and vomiting if eaten, so keep me away from children and pets.

PLANTS TO SPARK CREATIVITY

Some of our greatest inspirations come from nature, most of them from the awe of its beauty and intricacy. The individuality, complexity, form, and colour of plants open the mind to the wonders of creativity and the infinite possibilities of our imagination.

◄ Velvet plant
Gynura aurantiaca

With the ultraviolet purple hairs on its leaves, the gothic, almost supernatural look of the velvet plant will inspire you to think outside the box and not follow the norm – just like this unique plant. It will electrify your mind, spark your creativity, and pique your curiosity. *See pp.90–91*

◄ King begonias
Begonia rex cultivars

The intricately patterned leaves of these begonias look almost hand-painted, with the fusion of colours resembling our thought patterns as they lead to unexpected places and inspire new visions. The bold shapes of this plant will inspire bold ideas. *See pp.46–47.*

String of pearls ▸
Curio rowleyanus
The tumbling beads of this plant cascade down over the container like a waterfall, helping to evoke a feeling of calm and encouraging creative thoughts to flow. It's a plant that will leave you feeling stimulated and full of bright ideas, as calmness and creativity combine to help you innovate while remaining in a tranquil, composed state. *See pp.64–65*

◂ Moth orchid
Phalaenopsis
As its name suggests, the flowers of this orchid resemble a moth or butterfly frozen in time, and this image, in turn, can be likened to the spontaneity of thought. Just like a fluttering moth, our brains can go back and forth between ideas, stopping to capture the best ones. And just like a butterfly, our creativity can soar when this plant inspires. *See pp.112–113*

The prayer plant may surprise you with its rare dainty flowers

Prayer plant ▸
Maranta leuconeura '*Fascinator Tricolor*'
With a unique personality of its own, this plant mimics the daily human cycle by closing at night and reopening in the morning. It's a great reminder that, just like this plant, sometimes we need to step back, have a rest, then return to a great idea to refresh our minds. *See pp.104–105*

DWARF UMBRELLA TREE

Schefflera arboricola 'Compacta'

These wonderful indoor dwarf trees are sure to bring a touch of nature inside, and with them a sense of calm and serenity. The plant gets its common name from the shape of the foliage, which has groups of five to nine leaves that resemble jolly little umbrellas.

I'm symbolic

In feng shui, my leaves are said to capture positive energy and attract wealth, while the umbrella shape is believed to protect you from negative energies. This is one reason why you'll often see me in offices and restaurants.

{ **"**My umbrella leaves will **brighten** a rainy day**"** }

Help me thrive

 POSITION
Place in bright, filtered, or indirect light in temperatures above 10°C (50°F). Turn weekly to keep it growing straight rather than towards the light.

POTTING
Repot every 3 years in spring into the next pot size up. Use a well-drained potting mix, as these plants hate being overwatered.

GROWTH RATE
Will usually grow to heights of 1–2m (3¼–6½ft) tall inside a house, but can be pruned and trimmed to maintain the preferred height.

CARE
Do not overwater, especially in winter. Leave the top of the soil to dry out before watering. Feed monthly in summer with a balanced liquid fertilizer.

Look out for my flowers

I struggle to flower as a houseplant, but if I do you will spot red, pink, or white flowers that are shaped like tentacles, which explains one of my other common names, "octopus tree". Giving me as much sun as possible may encourage me to flower, and if I do, I will surprise you in the summer.

Give me a trim

I am a fast-growing plant, so I can become a little bit leggy. If I do become overgrown, I can be pruned back to give me a denser, healthier appearance. Once any weak stems are cut back to 7–10cm (2¾–4in), it will encourage new growth to appear in spring.

My compact flowers appear on clustered spikes

Let's make babies

I am an easy plant to take cuttings from. Just cut around 14cm (5½in) of a woody stem and insert it into an all-purpose houseplant compost where it will root. It's best to remove all of my old leaves as I will produce new ones, but remember not to plant me upside down.

BE AWARE: My sap can be harmful, so keep me away from children and pets.

PEACE LILY

Spathiphyllum 'Mauna Loa'

As its name suggests, the peace lily is full of positive, tranquil energy to cheer and relax you. It is a regular flowerer, easy to keep, and excellent at cleansing the air. The beautiful contrast between the deep glossy foliage and the white bracts will add a sophisticated and chic feeling to your room.

I'll remove toxins

With my broad leaves, I am a top purifying plant. I can help you keep a clean head by extracting high levels of benzene, formaldehyde, trichloroethylene, and ammonia from the atmosphere and giving you plenty of oxygen instead.

> **"I will encourage repose and cleanse the air"**

Help me thrive

 POSITION
Loves warm, humid conditions out of direct sunlight, so will grow well in shady corners if kept at 15°C (60°F) or above.

POTTING
Repot annually into the next size up if roots are growing through the bottom of the pot or circling inside. Best grown in soilless potting compost.

 GROWTH RATE
A vigorous yet compact grower, this plant can reach heights of 1m (3¼ft) and spread to 60cm (2ft), but it will grow more slowly in lower light.

CARE
Water when the top compost dries out and feed regularly with liquid houseplant fertilizer in summer. Water in winter to keep compost just moist.

Get to know me

My botanical name comes from the Greek *spath*, meaning "spathe", and *phyl*, which means "leaf", and references the fact that my spathe looks like a leaf. This is the white bract that protects my true flowers, which look like tiny bumps on my cream-coloured, tubular spadix.

Help me feel at home

I'm from South America and love warmth and high humidity. If I don't like the dry air my leaves may start to shrivel up. To avoid this, mist me regularly and/or place me on a tray of wet gravel.

I'm symbolic

I am known as the bringer of peace, because my white spathe represents a white flag, a common signal of truce. I'm also a great gift for someone going through difficult times, as I symbolize tranquillity.

Be mindful of me

I will flower freely and am tolerant of heavy shade. However, I may not flower in low light, as I need moderate light to photosynthesize productively, which in turn gives me the energy to bloom. On the other hand, if my spathes turn yellow, it may mean the light is too strong for me.

BE AWARE: I can be harmful if eaten, so keep me away from children and pets.

MADAGASCAR JASMINE

Stephanotis floribunda

I remember the first time I caught a whiff of the Madagascar jasmine's delightful scent – it made me smile and still does whenever I see one. It's a beautiful indoor climber, with glossy green leaves and delicate star-shaped flowers, often sold growing around a hoop as a living wreath.

My scent will relax you

My sweet-scented blooms will fill a room with fragrance, which can have a positive impact on your mood, helping to reduce stress, aid sleep, enhance physical and cognitive performance – and bring joy!

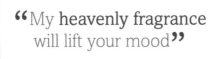

{ **"**My **heavenly fragrance** will lift your mood**"** }

Help me thrive

 POSITION
Grows well in good, indirect light, shaded from the hot sun. Needs temperatures of at least 17°C (63°F) in summer and 13°C (55°F) in winter.

 POTTING
Pot on into the next size up when roots grow through the bottom of the pot or begin to circle inside it. Use a well-draining compost.

 GROWTH RATE
This is a slow-growing climbing plant that, in a 20cm (8in) pot, will reach heights of 1–2m (3¼–6½ft) and a spread of 50cm (1½ft).

 CARE
Water regularly and feed every 2–3 weeks in the growing season. Reduce watering in winter but keep the air moist by misting through the year.

Get to know me

My horticultural name comes from the Greek *stephanos*, meaning "crown", and *otos*, meaning "ear", referring to the shape of my flowers and the pistils in my flowers, which look a little bit like ears. One of my common names is "bridal wreath", as my pristine white blooms are incredibly popular for wedding bouquets.

Let's make babies

Take 10cm (4in) cuttings from me in early summer, removing any leaves from the bottom 2.5cm (1in) of the stem. Dab the base of the stem into rooting hormone and pot up into well-drained compost. Cover with a cut plastic bottle to act as a cloche to keep humidity high.

Help me feel at home

Indoors, I can be displayed standing in a pot, in a hanging basket, or wandering up a support – I'm really flexible! Just position me in a spot where you can mist me regularly.

Help me flower

To keep me flowering, feed me with a high-potash liquid tomato feed every two or three weeks in spring and summer. You can also encourage my blooms by giving me a cool rest in winter, keeping me warm in the growing season, and pruning my stems after I flower.

CAPE PRIMROSE

Streptocarpus cultivars

The Cape primrose punches well above its weight in the houseplant world. Its delightful flowers will bring endless joy. And given its many wonderful types and colours, expanding your collection could turn into a satisfying hobby – which might even see you winning competitions for your prize blooms.

I'll make you happy

With my colourful trumpet-like blooms, I'll brighten up your day and make you feel happy. I can be found in purple, blue, pink, red, and white, or even in bi-coloured varieties with contrasting or darker veins, which add more depth to the colour of my flowers.

> **"My pretty blooms** will keep on coming**"**

Help me thrive

POSITION
Does well in good but filtered light, especially in spring and summer. Doesn't like temperatures much above 21°C (70°F).

POTTING
Flowers better if roots are growing close together so only repot in spring when the plant becomes rootbound. Use houseplant compost.

GROWTH RATE
Plants divide into 2 groups: rosettes can reach 25cm (10in) tall and 50cm (1½ft) wide; stemmed plants can reach 40cm (1¼ft) tall and 70cm (2¼ft) wide.

CARE
Water plants well from spring to autumn, letting the top soil dry out between waterings. Reduce watering in winter to keep the soil just moist.

Love my leaves

It's not just my flowers that will add style to your home; my long, velvety leaves will also bring attractive texture and colour. Please be kind to my leaves as they can tear easily – if they are accidentally torn, simply cut off the leaf, as I am quite forgiving and will grow new ones.

Get to know me

My name comes from the Greek word *strepos*, which means "twisted", and *karpos*, meaning "fruit". This refers to my long and twisted seed pods, but rather than allow me to use up my energy developing these seeds, pinch them off to encourage more flowers.

Keep me flowering

I can flower for most of the year in ideal conditions. Although I can flower in my natural habitat on the forest floor in South Africa, I still need enough light to photosynthesize, so I'll flower best by a bright window with indirect sunlight. Remove my dead flower stalks in spring or summer.

Let's make babies

In spring or early summer, pick new, healthy leaves from near my centre. Cut the leaves with a sharp knife into three or four sections, then insert the leaves into all-purpose compost the right way up, with cut edges in the mix so they can root. Water well, and babies will soon appear at the base of each leaf to be potted up separately.

PINK QUILL

Tillandsia cyanea

Originally from the rainforests of Ecuador, this energetic, almost neon-coloured plant brings a big wow to the home with its overlapping bright-pink bracts, which look like an old-fashioned quill and stay vibrant for months. Place it where you spend lots of time so you can enjoy it as much as possible.

I'll cheer you up

My bright pink quill is sure to grab your attention, with pink being a positive, exuberant, and cheering colour. My flowers, meanwhile, are a vivid purple, and this bold combination can't help but lift your spirits.

{ **"**My neon bracts will **electrify** your room**"** }

Help me thrive

 POSITION
Place in a warm room of 10–26°C (50–79°F), with bright, indirect light. Also likes good air circulation so a spot near a window will suit it.

 POTTING
Pot on to the next size up when roots grow through the bottom of the pot or circle inside it. Use a terrestrial bromeliad compost.

 GROWTH RATE
Grows quickly for its size to a height of 30cm (1ft) and a width of 40cm (1¼), taking around 3 years to reach its full flowering size.

 CARE
From late spring to mid-autumn, water when dry, giving a half-strength, low-nitrogen liquid feed monthly. Stop feeding in winter but keep watering.

Let's make babies

I am monocarpic, which means I die once I have flowered, but I can live on through the baby offsets that grow at my base. These "pups" can be separated from me and planted up in a fresh potting mix to grow on their own.

I'm sensitive
Please don't squeeze my quill as I store water there. If you squeeze me, I will quite literally cry, as tears will appear on the flat of my quill. I'm also sensitive to chemicals, so rainwater or distilled water is better for me than tap water.

Look out for my flowers
My vibrant quill may be as showy as a bunch of flowers but it will last a lot longer, as it's actually densely packed bracts, or modified leaves, that can live for up to three months. My real flowers are dainty purple blooms that appear from the side of my quill and are short-lived, only sticking around for a few days.

My violet flowers emerge briefly from between my bracts

Help me feel at home
I am an epiphyte, which means that in my natural rainforest habitat, I grow attached to trees, getting my moisture and nutrients from the humid air. This is why, when I'm kept as a houseplant, I will feel right at home if you mist me regularly as well as watering my soil.

SKY PLANT

Tillandsia spp.

These amazing little plants will instantly fill you with cheer and wonder. They survive on air alone so need no pot at all to grow in – they're clean and soil-free. They can go anywhere: on a shelf, stuck to driftwood, or even suspended in the air like a living mobile, offering a firework of fun foliage in the smallest spaces.

Get to know me

I have small scales on my leaves called trichomes that can absorb moisture from the air and even digest dust and turn it into food. My enchanting ornamental foliage is usually silvery-green, but some of my varieties can show hues of pink and purple as I mature.

> **"**I'm a **magical** plant that survives on air alone**"**

Help me thrive

 POSITION
Does well in a bright spot away from direct sunlight. A humid, well-ventilated place is ideal, as dry air will cause the leaf tips to turn brown.

 POTTING
Should not be grown in soil but hung in the air or mounted on objects for display. Will eventually form clumps, which can be split.

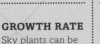 **GROWTH RATE**
Sky plants can be fast-growing, reaching up to 15–30cm (6in–1ft) tall and can produce offsets fairly quickly.

CARE
If grown in dry air, mist 2–3 times a week or give it a weekly soak (*see* Give me a soak). Feed monthly with an air plant foliar fertilizer.

Let's make babies

Once I have bloomed, I begin to deteriorate. However, I do grow offsets at my base that can be detached and mounted when they reach half my size, so you'll have an endless collection of my fascinating plantlets.

Help me feel at home

I'm native to Mexico and Costa Rica, so I can bring the tropics into your home – and because I'm an air plant you can get creative with how you display me. Just tuck me into a crevice in some wood, nestle me in a seashell, or attach me to a wire rack for a contemporary wall piece.

Look out for my flowers

When I'm mature, I will produce exotic colourful flowers – usually in pink and purple. My blooms should appear annually from my small spray of thin spiky leaves and can either be tiny, surprisingly large, or long and tubular, like a bursting firework.

Give me a soak

Water me well by plunging me into room-temperature water for about an hour once a week, then remove me and hold me upside down to ensure any excess water drips off. Alternatively, if I am mounted on a support, you can mist me two or three times a week until the excess water begins to drip off. It's best to use rainwater.

SILVER INCH PLANT

Tradescantia zebrina

Good looking, versatile, and tolerant of neglect, the silver inch plant is both mood-enhancing and a breeze to look after. With its pendulous trailing foliage, showing off the stunning variegation and purple and silver colouration, it makes an incredibly striking hanging-basket plant.

I'll liven up your room

I make an impressive feature when displayed in a hanging container, lifting the eye and livening up a dull corner. My shiny silver colouring mixed with green and purple tones will also lift your mood by softening hard corners and adding warmth to your room.

 " My beautiful **trailing foliage** will **dazzle you"**

Help me thrive

 POSITION
Loves full sun but will take some shade so can be placed in a south-, east- or west-facing room. Keep in a minimum temperature of 17°C (63°F).

POTTING
Repot small plants yearly in spring into the next pot size up, using an all-purpose potting mix. Larger plants benefit from a weekly feed instead.

GROWTH RATE
This is a vigorous, trailing grower – its stems can reach 50cm (1¾ft) long. Pinch tips off growing stems regularly to keep plants bushy.

CARE
Water to keep moist in summer. In winter, let the soil dry out first. Will tolerate feeding with every other watering or even once or twice a year.

Let's make babies

You can propagate me easily by taking an 8cm (3in) long cutting, removing the lower leaves so they aren't buried in the soil, and then potting up the stem in a standard houseplant potting mix. Water it and in about two or three weeks it will root.

Look out for my flowers

If you're lucky you may get a lovely surprise and catch a glimpse of my pinky-purple three-petalled flowers. They rarely appear on houseplants, but in my native country of Mexico they appear in clusters during spring or summer.

I bear three-petalled purple-pink flowers

Keep my colours bright

A position in a dimly lit spot will make the colours on my variegated leaves fade. To keep my silvers, greens, and purples looking brilliant, please keep me in a bright location with filtered light, making sure that my leaves don't get scorched by too much light.

BE AWARE: My sap is mildly toxic, so keep me away from children and pets, and wash your hands after handling me.

FLAMING SWORD

Vriesea splendens

It's hard to choose what I like most about this plant: the dominant bright sword of the bract and flower or the detailed variegation of the truly spectacular tiger-striped foliage. Together, these features make this plant a big wow for any home.

{ **"I dazzle** by day, help you **breathe** at night**"** }

I'll help you breathe

As a member of the bromeliad family, I breathe and produce oxygen at night, which is different to many other houseplants that only produce oxygen during the day. For this reason, I am a great bedroom plant, and when combined with other plants in different areas of the home, you will have 24-hour oxygen production.

Help me thrive

 POSITION
Best placed in a semi-shaded position, in temperatures of around 18°C (64°F) while actively growing and no less than 15°C (59°F) in winter.

 POTTING
Repot when roots grow through the bottom of the pot or circle on the inside. Use John Innes No.2 mixed with equal parts bark chips or perlite.

 GROWTH RATE
The arching leaves of this plant can reach 60cm (2ft) tall and can spread to a width of 45cm (1½ft). It is usually a slow grower.

CARE
Add water to the central cup and refresh every 2 weeks. Water the soil when dry. Add diluted liquid fertilizer to the cup monthly when in growth.

Let's make babies

I am monocarpic, so once I have bloomed, I slowly start to die, using my energy to produce new offsets instead. These plantlets can be potted up with a free-draining compost. Alternatively, attach them to some wood or driftwood by wrapping the root ball in coconut coir fibre, filling the bundle with a bit of free-draining potting mix, and then attaching the plant to the wood with thin wire.

Enjoy my "flame"

I get the name flaming sword from my long bract blade, which is a brilliant red, orange, or yellow colour and hosts my small tubular green-yellow flowers. My fiery-coloured bract can last for weeks.

My true flowers are small, yellow, and short-lived

My leaves add character

My foliage is just as striking as my flowers. The green and dark-purple tiger stripes of my splayed leaves make me ideal for a chic interior display. Look out for any brown patches appearing on them, as this may mean they are getting too much sun.

Help me feel at home

In my native South America, I use my roots to anchor myself to trees and rocks, and get nutrients from my central cup instead. This is why you need to water me by filling this cup with rain or distilled water – to replicate my natural home!

SPINELESS YUCCA

Yucca elephantipes

The yucca adds a refreshing pop of green and is a very easy-going house guest. It's hardy, tolerant of most places apart from a very chilly conservatory, and doesn't demand much! Its architectural trunk and palm-like leaves will be a bold, exotic presence in your home.

Bathe in my tropical beauty

My thick, woody stem and angular, sword-like foliage make a bold statement in a modern scheme without taking up too much floor space. And because I look like a miniature palm tree, you can take a tropical forest bath in my presence, helping you drift off to an island paradise.

> **"**I will transport you to a **tropical paradise"**

My bell flowers grow on spikes that shoot from my crown

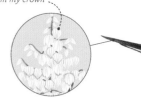

Help me thrive

 POSITION
Loves full sun but will tolerate light shade, although this will slow its growth and lighten the leaves. Turn occasionally so every part gets the sun.

 POTTING
Doesn't often require repotting as it likes crowded roots, but if roots grow out of the drainage holes, repot in the spring using a loam-based compost.

Let's make babies
You can grow my family in spring, either by potting up new offsets that appear from my trunk or by taking cuttings from my mature growth. Take a cutting of at least 8cm (3in) and remove all but the top leaves, then add some rooting hormone to the base and pot it up. Water the baby when the top of the soil dries out.

 **GROWTH RATE**
This slow-growing plant generally reaches 1m (3¼ft) tall and 30cm (1ft) wide after 5 years, although it can ultimately reach 1.8m (6ft).

 CARE
Water moderately in the summer, waiting until the top 5cm (2in) of soil has dried out in between waterings. Only water sparingly in winter.

Be mindful of my leaves

I'm known as the "spineless" yucca because although my leaves are pointed, they are not as razor sharp as my outdoor-growing cousins. I could still poke someone in the eye, though, so keep me out of reach of curious youngsters and pets.

Look out for my flowers

I'm generally chosen for my upright spiked leaves and statement trunks, so you may be surprised to know that I can produce flowers. This usually only happens when I'm in my native country of Mexico; my blooms rarely appear when I'm grown as a houseplant, but when they do, they are white bell-shaped flowers with a sweet scent.

Keep me grounded

My top-heavy appearance means I may be best planted in a deep and heavy container that will keep me firmly grounded and stop me from toppling over. An impressive plant like me needs an impressive pot, which will have a great impact in your home.

BE AWARE: I am harmful if eaten, so keep me away from children and pets.

INDEX

ABOUT THE AUTHOR
David Domoney, C Hort., FCI Hort.
David Domoney is one of Britain's leading experts on houseplants. A true believer in the power of plants to improve our mental and physical health, David is currently taking the nation by storm with a series of engaging talks on this subject, and supports a number of charities that share the same philosophy. David is Garden Champion for the mental health charity SANE, which seeks to support those living with mental illness. He is also an ambassador for Thrive, a charity that uses therapeutic horticulture to improve the lives of those living with physical disability, and a patron for Greenfingers charity, which creates tranquil garden spaces for families at children's hospices across the UK. He is also the founder of the British campaign Cultivation Street, supporting community gardens for better health.

David is a chartered horticulturalist and an ambassador for the Chartered Institute of Horticulture, a commercial board member for the Royal Horticultural Society, and a patron of the Birmingham Botanical Gardens. He has been awarded over 30 RHS medals for his garden designs, floral displays, and science and education exhibits, including Gold at the Chelsea Flower Show. Prince Edward personally selected David to receive the Excellence in Horticulture award in 2018.

As a presenter on Britain's most popular gardening TV show, ITV1's *Love Your Garden*, David has been appearing alongside Alan Titchmarsh before audiences of over four million for the last 10 years. For more than a decade he has also been the resident gardening presenter on ITV1's *This Morning*.

Acknowledgements
Author My thanks go to Danielle Walsh for her constant encouragement and all her research, support, and dedication in helping me achieve my deadlines and supervising the proofreading. Also, a big thanks to Rosie Irving for her research and proofreading, and to Charlotte Robertson, my literary agent, for her kind advice and encouragement. A big thank you to Alastair Laing and Holly Kyte at DK for their support and enthusiasm.

Thanks to my family: Adele, and our children Alice, Abigail, and Lance, who constantly remind me of the value of appreciating nature in this modern world, and my parents, Ray and Jean, for their encouragement and love as always.

Finally, a huge thank you to all the staff of indoor plant departments in garden centres for their dedication to encouraging everyone to enjoy growing houseplants.

Publisher DK would like to thank Simon Esdale at Flamingo Plants Ltd for his help acquiring plants and hosting the photoshoot; Sally Smallwood for photography assistance; Marie Lorimer for indexing.

Picture credits
The publisher would like to thank the following for their kind permission to reproduce their photographs:

(Key: a-above; b-below/bottom; c-centre; f-far; l-left; r-right; t-top)
94–95 123RF.com: stevanzz (b). **103 123RF. com:** stevanzz. **134–135 Dreamstime.com:** Kitithat Pansang (b). **144:** Heather Hayton (tl).

Cover images: *Back cover left to right*: Peter Anderson, Ruth Jenkinson, Rob Streeter; *Spine top to bottom*: Tim Winter, Matthew Ward; *Front*: Photo credit: Ruth Jenkinson.

All other images © Dorling Kindersley
For further information see: www.dkimages.com